When to Rob a Bank

W9-CPN-549

ALSO BY STEVEN D. LEVITT
& STEPHEN J. DUBNER

FREAKONOMICS

A Rogue Economist Explores
the Hidden Side of Everything

SUPERFREAKONOMICS

Global Cooling, Patriotic Prostitutes, and Why
Suicide Bombers Should Buy Life Insurance

SUPERFREAKONOMICS

The Deluxe, Super-Illustrated Edition

THINK LIKE A FREAK

The Authors of *Freakonomics* Offer to
Retrain Your Brain

ALSO BY STEPHEN J. DUBNER

TURBULENT SOULS

A Catholic Son's Return to His Jewish Family

Also published as

CHOOSING MY RELIGION

A Memoir of a Family Beyond Belief

CONFESSIONS OF A HERO-WORSHIPER

THE BOY WITH TWO BELLY BUTTONS

When to Rob a  Bank

. . . AND 131 MORE WARPED SUGGESTIONS AND WELL-INTENDED RANTS

Steven D. Levitt & Stephen J. Dubner

HARPER LUXE

An Imprint of HarperCollinsPublishers

Illustration on title page and half title pages, ©Justin Baker/ImageZoo/ Corbis. Illustrations, chapter openers: page 1, ©iStock.com/MrPlumo; page 6, ©iStock.com/kumdinpitak; page 42, ©iStock.com/bubaone; page 54, ©iStock.com/Kreatiw; page 100, ©iStock.com/ChrisGorgio; page 110, ©iStock.com/CSA-Images; page 148, ©iStock.com/mstay; page 182, ©iStock.com/ottoflick; page 206, ©iStock.com/Mr_Vector; page 248, ©iStock.com/Daniel Villeneuve; page 285, ©iStock.com/megamix; page 302, adapted from an illustration by ©iStock.com/ninochka; and page 337, ©iStock.com/doodlemachine. Illustration on page 246, ©iStock .com/ artishokcs. Photograph on page 117 by Stephen J. Dubner.

WHEN TO ROB A BANK. Copyright © 2015 by Steven D. Levitt and Dubner Productions, LLC. All rights reserved. Printed in the United States of America. No part of this book may be used or reproduced in any manner whatsoever without written permission except in the case of brief quotations embodied in critical articles and reviews. For information address HarperCollins Publishers, 195 Broadway, New York, NY 10007.

HarperCollins books may be purchased for educational, business, or sales promotional use. For information please e-mail the Special Markets Department at SPsales@harpercollins.com.

FIRST HARPERLUXE EDITION

HarperLuxe™ is a trademark of HarperCollins Publishers

Library of Congress Cataloging-in-Publication Data is available upon request.

ISBN: 978-0-06-239272-5

15 ID/RRD 10 9 8 7 6 5 4 3 2 1

We dedicate this book to our readers. We are perpetually wowed by your vigor and grateful for your attention.

Contents

When to
Rob a
Bank

What Do Blogs and Bottled Water Have in Common?

Ten years ago, as we were about to publish a book called *Freakonomics*, we decided to start a companion website. It was called, unimaginatively, Freakonomics.com. The site happened to offer a blogging function.

Levitt, who is always a few years behind, had never heard of a blog, much less read or written one. Dubner explained the idea. Levitt remained unconvinced.

"Let's just give it a try," Dubner said. It was so early in our partnership that Levitt hadn't yet come to understand that those six words were Dubner's way of getting him to do all sorts of things he never intended to do.

So we gave the blog a try. Here's the first post we wrote:

UNLEASHING OUR BABY

Every parent thinks he has the most beautiful baby in the world. Evolution, it seems, has molded our brains so that if you stare at your own baby's face day after day after day, it starts to look beautiful. When other people's children have food clotted on their faces, it looks disgusting; with your own kid, it's somehow endearing.

Well, we've been staring at the *Freakonomics* manuscript so much that it now looks beautiful to us—warts, clotted food, and all. So we started to think that maybe some people would actually want to read it, and after reading it, might even want to express their opinions about it. Thus, the birth of this website. We hope it's a happy (or at least happily contentious) home for some time to come.

And it *has* been a happy home! Our blog writing tends to be more casual, more personal, more opinionated than how we write our books; we are just as likely to float a question as to give a concrete answer. We've written things we only thought halfway through and later regretted. We've written things we *did* think through but also later regretted. But mostly, having the blog gave us good reason to stay curious and open about the world.

Unlike that first post, the vast majority of the blog entries were written by just one of us, not the pair, as in our book writing. We sometimes asked friends (and even enemies) to write for the blog; we've held "quorums" (asking a bunch of smart people to answer a tough question) and Q&As (with people like Daniel Kahneman and a high-end call girl named Allie). For several years, *The New York Times* hosted the blog, which gave it a veneer of legitimacy that wasn't quite warranted. But the *Times* eventually came to its senses and sent us off to do the thing we do, once more on our lonesome.

All these years, we routinely asked ourselves why we kept blogging. There was no obvious answer. It didn't pay; there wasn't any evidence the blog helped sell more copies of our books. In fact it may have cannibalized sales, since every day we were giving away our writing. But over time we realized why we kept at it: our readers liked reading the blog, and we loved our readers. Their curiosity and ingenuity and especially their playfulness have kept us at it, and in the pages to follow you will see ample evidence of their spirit.

Occasionally a reader would suggest that we turn our blog writing into a book. This struck us as a colossally dumb idea—until, one day not long ago, it didn't. What changed? Dubner was dropping off one of his kids at summer camp in Maine. In the middle of nowhere, they came upon a huge Poland Spring water-bottling plant. Having grown up in the middle of nowhere himself, Dubner had always thought it strange that so many people would pay good money for a bottle of water. And yet they do, to the tune of roughly $100 billion a year.

Suddenly a book of blog posts didn't seem so dumb. So in the tradition of Poland Spring, Evian, and other hydro-geniuses, we've decided to bottle something that was freely available and charge you money for it.

To be fair, we did go to the trouble of reading through the whole blog and picking out the best material. (It was gratifying to find that among eight thousand mostly mediocre posts, we actually had some good ones.) We edited and updated the posts as necessary, arranging them into chapters that make book sense. Chapter One, for instance, "We Were Only Trying to Help," addresses the abolition of academic tenure, alternatives to democracy, and how to think like a terrorist. "Limberhand the Masturbator and the Perils of Wayne" is about names that are strange, fitting, or strangely fitting. "When You're a Jet . . ." shows that once you start thinking like an economist, it's hard to turn it off—whether the subject is baby formula, animated films, or rancid chicken. Along the way, you will learn more than you ever wanted to know about our personal obsessions like golf, gambling, and the dreaded penny.

It has given us gargantuan pleasure over the years to put our warped thoughts into writing. We hope you enjoy peeking inside our heads to see what it's like to view the world through Freakonomics-colored glasses.

Chapter 1
We Were Only Trying to Help

Some of the best ideas in history—nearly all of them, in fact—sounded crazy at first. That said, a lot of crazy-sounding ideas truly are crazy. But how do you find out? One of the best things about having a blog is that you've got a place to run your craziest ideas up the flagpole and see just how quickly they get shot down. Of all the posts we've ever written, the first one in this chapter generated the quickest, loudest, angriest response.

IF YOU WERE A TERRORIST, HOW WOULD YOU ATTACK?

(SDL)

The TSA recently announced that most airplane carry-on restrictions will stay in place for now, although the ban has now been lifted on cigarette lighters. While it seems crazy to keep people from bringing toothpaste, deodorant, or water through security, it didn't seem so strange to ban lighters. I wonder whether the lighter manufacturers were lobbying for or against this rule change. On the one hand, having twenty-two thousand lighters confiscated per day would seem good for business; on the other hand, maybe fewer people will buy lighters if they can't travel with them.

Hearing about these rules got me thinking about what I would do to maximize terror if I were a terrorist with limited resources. I'd start by thinking about what really inspires fear. One thing that scares people is the thought that they could be a victim of an attack. With that in mind, I'd want to do something that everybody thinks might be directed at them, even if the individual probability of harm is very low.

Humans tend to overestimate small probabilities, so the fear generated by an act of terrorism is greatly disproportionate to the actual risk.

Also, I'd want to create the feeling that an army of terrorists exists, which I'd accomplish by pulling off multiple attacks at once, and then following them up with more shortly thereafter.

Third, unless terrorists always insist on suicide missions (which I can't imagine they would), it would be optimal to hatch a plan in which your terrorists aren't killed or caught in the act, if possible.

Fourth, I think it makes sense to try to stop commerce, since a commerce breakdown gives people more free time to think about how scared they are.

Fifth, if you really want to impose pain on the U.S., the act has to be something that prompts the government to pass a bundle of very costly laws that stay in place long after they have served their purpose (assuming they had a purpose in the first place).

My general view of the world is that simpler is better. My guess is that this thinking applies to terrorism as well. In that spirit, the best terrorist plan I have heard is one that my father thought up after the D.C. snipers created havoc in 2002. The basic idea is to arm twenty terrorists with rifles and cars, and arrange to have them begin shooting randomly at pre-set times all across the country. Big cities, little cities, suburbs, etc. Have them move around a lot. No one will know when and where the next attack will be. The chaos would be unbelievable, especially considering how few resources it would require of the terrorists. It would also be extremely hard to catch these guys. The damage wouldn't be as extreme as detonating a nuclear bomb in New York City, of course, but it sure would be a lot easier to obtain a handful of guns than a nuclear weapon.

I'm sure many readers have far better ideas. I would love to hear them. Consider that posting them on this blog could be a form of public service: I presume that a lot more folks who oppose and fight terror read this blog than actual terrorists. So by getting these ideas out in the open, it gives terror fighters a chance to consider and plan for these scenarios before they occur.

This post was published on August 8, 2007, the day the Freakonomics blog took up residence on The New York Times*'s website. That same day, in an interview with* The New York Observer, *Dubner was asked to explain why Freakonomics was the first outside blog that the* Times *had decided to publish. His answer reflected the fact that he used to work at the newspaper and knew well its standards and mores: "They know I'm not going to issue some sort of fatwa on the blog." As it turns out, Levitt's post soliciting ideas for a terrorist strike was considered exactly that. It drew so much heated response that the* Times *shut down the comments section after a few hundred comments. Here is a typical one: "You have got be kidding me. Ideas for terrorists? Think you are being cute? Clever? You are an idiot." This led Levitt to try again, the following day:*

TERRORISM, PART II
(SDL)

On the very first day that our blog was hosted by *The New York Times,* I wrote a post that generated the most hate mail I've gotten since the abortion-crime story first broke almost a decade ago. The people e-mailing me can't decide whether I am a moron, a traitor, or both. Let me try again.

A lot of the angry responses make me wonder what everyday Americans think terrorists do all day. My guess is that they brainstorm ideas for terrorist plots. And you have to believe that terrorists are total idiots if it never occurred to them after the Washington, D.C., sniper shootings that maybe a sniper plot wasn't a bad idea.

The point is this: there is a virtually infinite array of incredibly simple strategies available to terrorists. The fact that it has been six years since the last major terrorist attack in the United States suggests either that the terrorists are incompetent, or that perhaps their goal isn't really to generate terror. (A separate factor is the prevention efforts by law enforcement and the government; I'll address that later.)

Many of the angry e-mails I received demanded that I write a post explaining how we stop terrorists. But the obvious answer is a disappointing one: if terrorists want to engage in low-grade, low-tech terror, we are powerless to stop it.

That is the situation in Iraq right now, and, to a lesser degree, in Israel. That was also more or less the situation with the IRA a while back.

So what can we do? Like the British and Israelis have done, if faced with this situation, Americans would figure out how to live with it. The actual cost of this low-grade terrorism in terms of human lives is relatively small, compared to other causes of death like motor-vehicle crashes, heart attacks, homicide, and suicide. It is the fear that imposes the real cost.

But just as people in countries with runaway inflation learn relatively quickly to live with it, the same happens with terrorism. The actual risk of dying from an attack while riding a bus in Israel is low—and so, as Gary Becker and Yona Rubinstein have shown, people who have a lot of experience riding Israeli buses don't respond much to the threat of bombings. Similarly, there is little wage premium for being a bus driver in Israel.

Beyond this, I think there are a few more prospective things we can do. If the threat is from abroad, then we can do a good job screening risky people from entering the country. That, too, is obvious. Perhaps less obvious is that we can do a good job following potential risks after they enter the country. If someone enters on a student visa and isn't enrolled in school, for instance, he is worth keeping under close surveillance.

Another option is one the British have used: putting cameras everywhere. This is very anti-American, so it probably would never fly here. I also am not sure it is a good investment. But the recent terrorist attacks in the U.K. suggest that these cameras are at least useful after the fact in identifying the perpetrators.

The work of my University of Chicago colleague Robert Pape suggests that the strongest predictor of terrorist acts is the occupation of a group's territory. From that perspective, having American troops in Iraq is probably not helping to reduce terrorism—although it may be serving other purposes.

Ultimately, though, it strikes me that there are two possible interpretations of our current situation vis-à-vis terrorism.

One view is this: the main reason we aren't currently being decimated by terrorists is that the government's anti-terror efforts have been successful.

The alternative interpretation is that the terror risk just isn't that high and we are greatly overspending on fighting it, or at least appearing to fight it. For most government officials, there is much more pressure to look like you are trying to stop terrorism than there is to actually stop it. The head of the TSA can't be blamed if a plane gets shot down by a shoulder-launched missile, but he is in serious trouble if a tube of explosive toothpaste takes down a plane. Consequently, we put much more effort into the toothpaste even though it is probably a much less important threat.

Likewise, an individual at the CIA isn't in trouble if a terrorist attack happens; he or she is only in trouble if there is no written report that details the possibility of such an attack, which someone else should have followed up on, but never did because there are so many such reports written.

My guess is that the second scenario—the terrorism threat just isn't that great—is the more likely one. Which, if you think about it, is an optimistic view of the world. But that probably still makes me a moron, a traitor, or both.

HOW ABOUT A "WAR ON TAXES"?
(SJD)

David Cay Johnston, who does an incredible job covering U.S. tax policy and other business issues for *The New York Times*, reports that the IRS is outsourcing the collection of back taxes to third parties, a.k.a. collection agencies. "The private debt collection program is expected to bring in $1.4 billion over 10 years," he writes, "with the collection agencies keeping about $330 million of that, or 22 to 24 cents on the dollar."

Maybe that seems like too big a cut to surrender. And maybe people will be worried about the collection agencies having access to their financial records. But what's most striking to me is that the IRS knows who owes the money and knows where to find it, but because it is understaffed cannot afford to collect it. So it has to hire someone else to do it, at a stiff price.

The IRS admits that external collection is far more expensive than internal collection. Former commissioner Charles O. Rossotti once told Congress that if the IRS hired more agents, it "could collect more than $9 billion each year and spend only $296 million—or about three cents on the dollar—to do so," Johnston writes.

Even if Rossotti was exaggerating by a factor of five, the government would still be getting a better deal by hiring more agents than by contracting to a third party that takes a 22 percent cut. But Congress, which oversees the IRS budget, is famously reluctant to give the agency more resources to do its job. We touched on this subject in a *Times* column of our own:

> A main task of any IRS commissioner . . . is to beg Congress and the White House for resources. For all the obvious appeal of having the IRS collect every dollar owed to the government, it is just as obviously unappealing for most politicians to advocate a more vigorous IRS. Michael Dukakis tried this during his 1988 presidential campaign, and—well, it didn't work.

Left to enforce a tax code no one likes upon a public that knows it can practically cheat at will, the IRS does its best to fiddle around the edges.

Why does Congress act as it does? Maybe our congressmen are a bunch of history buffs so imbued with the spirit of our republic that they remember the Boston Tea Party too well and are scared of how the populace might revolt if they ramp up tax enforcement. But keep in mind that we are talking about tax enforcement here, which is the IRS's job, and not tax law, which is Congress's responsibility. In other words, Congress is happy to set the tax rates that it does; but it doesn't want to be seen as giving too much comfort to the bad cops who have to go out and collect those tax dollars.

So maybe they need to relabel their effort to get all the tax money that is owed. Since Congress approves so much money for the War on Terror and the War on Drugs, maybe it's time for them to launch a War on Taxes—well, really, a War on Tax Cheats. What if they could demonize the tax cheats so thoroughly, emphasizing that the "tax gap" (the difference between taxes owed and money collected) is about the size of the federal deficit: Would that make it more politically palatable to give the IRS the resources to collect the money that is owed? Maybe they could put pictures of tax cheats on milk cartons, on flyers at the post office, even on *America's Most Wanted*. Would that do the trick? Would a properly managed War on Tax Cheats fix the problem?

For now, we'll have to settle for the IRS turning over the job to collection agencies who will collect some money but not nearly as much as is owed. Which means a lot of money—a lot of tax money, that is, collected from the people who don't cheat—continuing to go down the drain.

IF PUBLIC LIBRARIES DIDN'T EXIST, COULD YOU START ONE TODAY?
(SJD)

Raise your hand if you hate libraries.

Yeah, I didn't think so. Who could possibly hate libraries?

Here's one suggestion: book publishers. I am probably wrong on this, but if you care about books, hear me out.

I had lunch recently with a few publishing folks. One of them had just returned from a national librarians' conference, where it was her job to sell her line of books to as many librarians as possible. She said that there were twenty thousand librarians in attendance; she also said that if she got one big library system, like Chicago's or New York's, to buy a book, that could mean a sale of as many as a few hundred copies, since many library branches carry several copies of each book.

That sounds great, doesn't it?

Well . . . maybe not. Among writers, there is a common lament. Someone comes up to you at a book signing and says, "Oh, I loved your book so much, I got it from the library and then told all my friends to go to the library, too!" And the writer thinks, "Gee, thanks, but why didn't you buy it?"

The library bought its copy, of course. But let's say fifty people will read that copy over the life of the book. If the library copy hadn't existed, surely not all fifty of those people would have bought the book. But imagine that even five people would have. That's four additional book sales lost by the writer and the publisher.

There's another way to look at it, of course. Beyond the copies that libraries buy, you could argue that, in the long run, libraries augment overall book sales along at least a few channels:

1. Libraries help train young people to be readers; when those readers are older, they buy books.

2. Libraries expose readers to works by authors they wouldn't have otherwise read; readers may then buy other works by the same author, or even the same book to have in their collection.

3. Libraries help foster a general culture of reading; without it, there would be less discussion, criti-cism, and coverage of books in general, which would result in fewer book sales.

But here's the point I'm getting to: if there were no such thing today as the public library and someone like Bill Gates proposed to establish them in cities and towns across the U.S. (much like Andrew Carnegie once did), what would happen?

I am guessing there would be a huge pushback from book publishers. Given the current state of debate about intellectual property, can you imagine modern publishers being willing to sell one copy of a book and then have the owner let an unlimited number of strang-ers borrow it?

I don't think so. Perhaps they'd come up with a licensing agreement: the book costs twenty dollars to own, with an additional two dollars per year for every year beyond year one it's in circulation. I'm sure there would be a lot of other potential arrangements. And

I am just as sure that, like a lot of systems that evolve over time, the library system is one that, if it were being built from scratch today, wouldn't look anything like it actually does.

LET'S JUST GET RID OF TENURE (INCLUDING MINE)
(SDL)

If there was ever a time when it made sense for economics professors to be given tenure, that time has surely passed. The same is likely true of other university disciplines, and probably even more true for high-school and elementary-school teachers.

What does tenure do? It distorts people's effort so that they face strong incentives early in their career (and presumably work very hard early on as a consequence) and very weak incentives forever after (and presumably work much less hard on average as a consequence).

One could imagine some models in which this incentive structure makes sense. For instance, if one needs to learn a lot of information to become competent, but once one has the knowledge it does not fade, and effort is not very important. That model may be a good description of learning to ride a bike, but it is a terrible model of academics.

From a social standpoint, it seems like a bad idea to make incentives so weak after tenure. Schools get stuck with employees who are doing nothing (at least not doing what they are presumably being paid to do). It also is probably a bad idea to give such strong incentives pre-tenure—even without tenure, young faculty have lots of reasons to work hard to build a good career.

The idea that tenure protects scholars who are doing politically unpopular work strikes me as ludicrous. While I can imagine a situation where this issue might arise, I am hard-pressed to think of actual cases where it has been relevant. Tenure does an outstanding job of protecting scholars who do no work or terrible work, but is there anything in economics which is high quality but so controversial that it would lead to a scholar being fired? Anyway, that is what markets are for. If one institution fires an academic primarily because they don't like his or her politics or approach, there will be other schools happy to make the hire. There are, for instance, cases in recent years in economics where scholars have made up data, embezzled funds, etc., but still have found good jobs afterward.

One hidden benefit of tenure is that it works as a commitment device to get departments to fire mediocre people. The cost of not firing at a tenure review is higher with tenure in place than it is without it. If it is painful to fire people, without tenure the path of least resistance may be to always say you will fire the person the next year, but never do it.

Imagine a setting where you care about performance (e.g., a professional football team, or as a currency trader). You wouldn't think of granting tenure. So why do it in academics?

The best-case scenario would be if all schools could coordinate on dumping tenure simultaneously. Maybe departments would give the deadwood a year or two to prove they deserved a slot before firing them. Some non-producers would leave or be fired. The rest of the tenure-age economists would start working harder. My guess is that salaries and job mobility would not change that much.

Absent all schools moving together to get rid of tenure, what if one school chose to unilaterally revoke tenure? It seems to me that it might work out just fine for that school. It would have to pay the faculty a little extra to stay in a department without an insurance policy in the form of tenure. Importantly, though, the value of tenure is inversely related to how good you are. If you are way over the bar, you face almost no risk if tenure is abolished. So the really good people would require very small salary increases to compensate for no tenure, whereas the really bad, unproductive economists would need a much bigger subsidy to remain in a department with tenure gone. This works out fantastically well for the university because all the bad people end up leaving, the good people stay, and other good people from different institutions want to come to take advantage of the salary increase at the tenure-less school. If the U of C told me that they were going to revoke my tenure, but add $15,000 to my salary, I would be happy to take that trade. I'm sure many others would as well. By dumping one unproductive, previously tenured faculty member, the university could compensate ten others with the savings.

WHY DON'T FLIGHT ATTENDANTS GET TIPPED?
(SJD)

Think of all the service people who habitually get tips: hotel bellmen, taxi drivers, waiters and waitresses, the guys who handle curbside baggage at airports, sometimes even the baristas at Starbucks. But not flight attendants. Why not?

Maybe it's because they're thought to earn a pretty good living and don't need the tips. Maybe it's because they're simply thought to be salaried employees of a sort that for whatever reason shouldn't accept tips. Maybe for some reason they are actually prohibited from accepting tips. Maybe it harks back to the day when most flight attendants were women and most passengers were men—and given the somewhat mystical (or perhaps mythical) reputation of the amorous businessman and the foxy stewardess, the exchange of money at flight's end may have raised some questions about just what the stewardess had done to deserve the tip.

Still, it's very odd to me that so many service people who perform similar functions get tipped and that flight attendants don't. Especially when they often work so hard for so many people, running back and forth with drinks, pillows, headphones, etc. Yes, I know that most people are pretty unhappy with the airline experience these days, and I know that the occasional flight attendant is crabby beyond belief, but in my experience most of them do a really great job, often under trying circumstances.

It's not that I'm advocating for yet another kind of worker to get tips. But having flown a lot lately, and seeing how hard flight attendants work, it struck me as odd that they don't get tipped. At least I've never seen anyone tip a flight attendant. And when I asked flight attendants on my last five flights if they'd ever gotten a tip, each of them said no, never. Their reactions to my question ranged from quizzical to hopeful.

I think on my flight home today, I'll simply slip the tip instead of asking the question, and see what happens.

Update: I tried, and failed. "A flight attendant is not a waitress," I was told—so forcefully that I felt terrible for even trying to put money in the woman's hands.

WANT TO FIX NEW YORK AIR CONGESTION? SHUT DOWN LAGUARDIA (SJD)

The Department of Transportation just canceled its plan to auction off landing and takeoff slots at New York City's three airports. The idea was to use market forces to ease congestion, but in the face of industry backlash (and legal threats), transportation secretary Ray LaHood called off the auction.

"We're still serious about tackling aviation congestion in the New York region," LaHood says. "I'll be talking with airline, airport, and consumer stakeholders, as well as elected officials, over the summer about the best ways to move forward."

The three major airports serving New York— J.F.K., Newark-Liberty, and LaGuardia—are famously high-ranked when it comes to congestion and delays. And since so many flights elsewhere connect through New York, their delays affect air traffic everywhere.

During a recent ground delay at LaGuardia, I got to talking with an off-duty pilot for a major airline who was extraordinarily knowledgeable about every single airline question I could think to ask him. When I asked for his take on New York air congestion, he said the solution was easy: shut down LaGuardia.

The problem, as he explained it, is that the airspace for each of the three airports extends cylindrically into the sky above its ground position. Because of their relative proximity, the three airspace cylinders affect one another significantly, which creates congestion not just because of volume but because pilots have to thread the needle and fly unnecessarily intricate approach routes in order to comply.

If the LaGuardia cylinder were eliminated, he said, Newark and J.F.K. would both operate much more freely—and since LaGuardia handles far less traffic than the other two airports, it is the obvious choice for shuttering.

But there's a problem: LaGuardia is the favored airport of the people with the most political power in New York, since it is a very short ride from Manhattan. So it's unlikely to happen, at least anytime soon. But if it did, my new pilot friend insisted, New York air travel would move from nightmare to dream.

I have to admit that LaGuardia is my favorite airport, since I live in Manhattan and can usually get there in about fifteen minutes. On every other dimension, meanwhile, it is less pleasant and comfortable than either Newark or J.F.K.

That said, if eliminating LaGuardia had the cascading effect of streamlining all New York air traffic, I would personally help start knocking it down. Let's say that I, and every other New York traveler, spend an average of thirty wasted minutes on every inbound and outbound flight in any of the three airports. (That's probably being generous.) That's a one-hour delay for each round trip. If I had to go to Newark or J.F.K. for every flight, I'd spend a little bit less than an extra hour on a round-trip ground commute to the airport—so with no delays, I'd at least be breaking even. Everyone who lived closer to either airport would obviously do even better. And then we'd get to start adding up all the time and productivity regained around the country by eliminating the inevitable New York airport delays.

WHY RESTORING THE MILITARY DRAFT IS A TERRIBLE IDEA
(SDL)

A long report in *Time* magazine carries the headline "Restoring the Draft: No Panacea."

Milton Friedman must be turning over his grave at the mere suggestion of a military draft. If the problem is that not enough young people are volunteering to fight in Iraq, there are two reasonable solutions: 1) take the troops out of Iraq; or 2) compensate soldiers well enough that they are willing to enlist.

The idea that a draft presents a reasonable solution is completely backward. First, it puts the "wrong" people in the military—people who are either uninterested in a military life, not well equipped for one, or who put a very high value on doing something else. From an economic perspective, those are all decent reasons for not wanting to be in the military. (I understand that there are other perspectives—for example, a sense of debt or duty to one's country—but if a person feels that way, it will be factored into his or her interest in military life.)

One thing markets are good at is allocating people to tasks. They accomplish this through wages. As such, we should pay U.S. soldiers a fair wage to compensate them for the risks they take! A draft is essentially a large, very concentrated tax on those who are drafted. Economic theory tells us that is an extremely inefficient way to accomplish our goal.

Critics might argue that sending economically disadvantaged kids to die in Iraq is inherently unfair. While I wouldn't disagree that it's unfair that some people are born rich and others poor, given that income disparity exists in this country, you'd have to possess a low opinion of the decision-making ability of military enlistees to say that a draft makes more sense than a volunteer army. Given the options they face, the men and women joining the military are choosing that option over the others available to them. A draft may make sense as an attempt to reduce inequality; but in a world filled with inequality, letting people choose their own paths is better than dictating one for them. As a perfect example of this, the Army is currently offering $20,000 "quick ship" bonuses to those who are willing to ship out to basic training within thirty days of signing up. (This bonus likely has something to do with the fact that the Army just hit its monthly recruiting goal for the first time in a while.)

It would be even better if the government was required to pay fair wages to soldiers during wartime— i.e., if combat pay was market-determined and soldiers could opt to leave whenever they wanted, like most jobs. If that were the case, the cost to the government would skyrocket and more accurately reflect the true costs of war, leading to a truer assessment of whether the benefits of military action outweigh the costs.

Critics also argue that if there were more affluent Caucasians in the military, we wouldn't be in Iraq. That is probably true, but it doesn't automatically mean that a draft is a good idea. A draft would make fighting wars much less efficient, which should mean fewer wars. But it may be the case that, if you can fight a war efficiently, it is worth fighting—even if it's not worth fighting inefficiently. To be clear, I am not saying this particular war is necessarily worth fighting—just that, in theory, this could be true.

As a side point, the current system of relying on reservists doesn't seem like a good one either. Essentially, it involves the government overpaying reservists when they aren't needed, and underpaying them when they are needed. This setup shifts all the risk from the government to the reservists. From an economic perspective, such a result doesn't make any sense, because individuals shouldn't/don't like risk. Ideally, you would want a system in which the payment to reservists is extremely low in peacetime, and high enough in wartime that they would be indifferent to being called up or not.

A FREAKONOMICS PROPOSAL TO HELP THE BRITISH NATIONAL HEALTH SERVICE
(SDL)

In the first chapter of our book *Think Like a Freak,* we recount an ill-fated interaction that Dubner and I had with David Cameron shortly before he was elected prime minister of the U.K. (In a nutshell, we joked with Cameron about applying the same principles he espoused for health care to automobiles; it turns out you don't joke with prime ministers!)

That story has riled up some people, including an economics blogger named Noah Smith, who rails on us and defends the NHS.

I should start by saying I have nothing in particular against the NHS, and I also would be the last one to ever defend the U.S. system. Anyone who has ever heard me talk about Obamacare knows I am no fan of it, and I never have been.

But it doesn't take a whole lot of smarts or a whole lot of blind faith in markets to recognize that when you don't charge people for things (including health care), they will consume too much of it. I guarantee you that if Americans had to pay out of their own pockets the crazy prices that hospitals charge for services, a much smaller share of U.S. GDP would go to health care. And, of course, the same would be true in the U.K.

Smith ends his critique by writing:

But I don't think Levitt has a model. What he has is a simple message ("all markets are the same"), and a strong prior belief in that message.

Smith could not have known, based on what's in *Think Like a Freak*, that we actually do have a model for the NHS. And, indeed, I proposed the model to Cameron's team after he left the meeting.

If nothing else, the model is admirably simple.

On January 1 of each year, the British government would mail a check for £1,000 to every British resident. They can do whatever they want with that money, but if they are being prudent, they might want to set it aside to cover out-of-pocket health care costs. In my system, individuals are now required to pay out of pocket for 100 percent of their health care costs up to £2,000, and 50 percent of the costs between £2,000 and £8,000. The government pays for all expenses over £8,000 in a year.

From a citizen's perspective, the best-case scenario is that they use no health care, so they end up £1,000 to the positive. Well over half of U.K. residents will end up spending less than £1,000 on health care in a given year. The worst case for an individual is that he/she ends up consuming more than £8,000 of health care, so that he/she ends up £4,000 in the red (he/she spends £5,000 on health care, but this is offset by the £1,000 gift at the beginning of the year).

If it turns out that consumers are sensitive to prices (i.e., that the most basic principle of economics holds, and demand curves slope downward), total spending on health care will decrease. In simulations we've run at the Greatest Good, we estimate that total health care costs might decline by roughly 15 percent. That is a decrease in spending of nearly twenty billion pounds. This decrease comes because a) competition will likely lead to increased efficiency; and b) consumers will cut out the low-value health care services they are currently using only because the services come for free.

Everyone remains protected against catastrophic illness.

Like any government program, there are winners and losers. The majority of Brits will be better off in the scenario I laid out, but those who need to spend a lot on health care in a particular year will be worse off. That is because the system I propose provides only partial insurance—which retains incentives for consumers to make prudent choices. The health care system would then mimic the rest of life. When my TV breaks, I have to buy a new one. I'm worse off than the guy whose TV did not break. When my roof needs to be replaced, it's expensive, and I'm worse off than if the roof didn't need replacement. There's nothing immoral about this; it is just the way the world usually works.

There are, no doubt, many improvements that could be made to this simple proposal. For instance, maybe the cash payment to the elderly at the beginning of the year should be larger than that to those who are younger. Maybe the cash payment is bigger to those who have chronic illnesses, etc.

I have no idea whether this sort of plan could be politically viable, but I have done some informal polling of the British electorate. Every time I take a cab in London, I ask my driver whether he would be in favor of my proposal. Probably the cabbies are just being polite, but roughly 75 percent of them say they would prefer my plan to the current system.

Perhaps, then, it is time for another audience with the prime minister . . .

AN ALTERNATIVE TO DEMOCRACY?
(SDL)

With the U.S. presidential election nearly here, everyone seems to have politics on their mind. Unlike most people, economists tend to have an indifference toward voting. The way economists see it, the chances of an individual's vote influencing an election outcome is vanishingly small, so unless it is fun to vote, it doesn't make much sense to do so. On top of that, there are a number of theoretical results, most famously Arrow's Impossibility Theorem, which highlight how difficult it is to design political systems/voting mechanisms that reliably aggregate the preferences of the electorate.

Mostly, these theoretical explorations into the virtues and vices of democracy leave me yawning.

Last spring, however, my colleague Glen Weyl mentioned an idea along these lines that was so simple and elegant that I was amazed no one had ever thought of it before. In Glen's voting mechanism, every voter can vote as many times as he or she likes. The catch, however, is that you have to pay each time you vote, and the amount you have to pay is a function of the square of the number of votes you cast. As a consequence, each extra vote you cast costs more than the previous vote. Just for the sake of argument, let's say the first vote costs you $1. Then to vote a second time would cost $4. The third vote would be $9, the fourth $16, and so on. One hundred votes would cost you $10,000. So eventually, no matter how much you like a candidate, you choose to vote a finite number of times.

What is so special about this voting scheme? People end up voting in proportion to how much they care about the election outcome. The system captures not just which candidate you prefer, but how strong your preferences are. Given Glen's assumptions, this turns out to be Pareto efficient—i.e., no person in society can be made better off without making someone else worse off.

The first criticism you'll likely make against this sort of scheme is that it favors the rich. At one level that is true relative to our current system. It might not be a popular argument, but one thing an economist might say is that the rich consume more of everything—why shouldn't they consume more political influence? In our existing system of campaign contributions, there can be little doubt that the rich already have far more influence than the poor. So restricting campaign spending, in conjunction with this voting scheme, might be more democratic than our current system.

Another possible criticism of Glen's idea is that it leads to very strong incentives for cheating through vote buying. It is much cheaper to buy the first votes of a lot of uninterested citizens than it is to pay the price for my one-hundredth vote. Once we put dollar values on votes, it is more likely that people will view votes through the lens of a financial transaction and be willing to buy and sell them.

Given we've been doing "one person, one vote" for so long, I think it is highly unlikely that we will ever see Glen's idea put into practice in major political elections. Two other economists, Jacob Goeree and Jingjing Zhang, have been exploring a similar idea to Glen's and testing it in a laboratory environment. Not only does it work well, but when given a choice between standard voting and this bid system, the participants usually choose the bid system.

This voting scheme can work in any situation where there are multiple people trying to choose between two alternatives—e.g., a group of people trying to decide which movie or restaurant to go to, housemates trying to decide which of two TVs to buy, etc. In settings like those, the pool of money that is collected from people voting would be divided equally and then redistributed to the participants.

My hope is that a few of you might be inspired to give this sort of voting scheme a try. If you do, I definitely want to hear about how it works out!

WOULD PAYING POLITICIANS MORE ATTRACT BETTER POLITICIANS?
(SJD)

Whenever you look at a political system and find it wanting, one tempting thought is this: maybe we have subpar politicians because the job simply isn't attracting the right people. And, therefore, if we were to significantly raise politicians' salaries, we would attract a better class of politician.

This is an unpopular argument for various reasons, one of them being that it would be the politicians themselves who have to lobby for higher salaries, and that isn't politically feasible (especially in a poor economy). Can you imagine the headlines?

But the idea remains attractive, doesn't it? The idea is that by raising the salaries of elected and other government officials, you would a) signal the true importance of the job; b) attract a kind of competent person who might otherwise enter a more remunerative field; c) allow politicians to focus more on the task at hand rather than worry about their income; and d) make politicians less susceptible to the influence of moneyed interests.

Some countries already pay their government officials a lot of money—Singapore, for instance. From Wikipedia:

> Ministers in Singapore are the highest paid politicians in the world, receiving a 60% salary raise in 2007 and as a result Prime Minister Lee Hsien Loong's pay jumped to S$3.1 million, five times the US$400,000 earned by President Barack Obama. Although there was a brief public outcry regarding the high salary in comparison to the size of the country governed, the government's firm stance was that this raise was required to ensure the continued efficiency and corruption-free status of Singapore's "world-class" government.

Although Singapore recently cut its politicians' pay substantially, the salaries remain relatively very high.

But is there any evidence that paying politicians more actually improves quality? A research paper by Claudio Ferraz and Frederico Finan argues that it did for municipal governments in Brazil:

> Our main findings show that [paying a] higher wage increases political competition and improves the quality of legislators, as measured by education, type of previous profession, and political experience in office. In addition to this positive selection, we find that wages also affect politicians' performance, which is consistent with a behavioral response to a higher value of holding office.

Another, more recent paper by Finan, Ernesto Dal Bó, and Martin Rossi finds that the quality of civil servants also improves when they are paid more, this time in Mexican cities:

> We find that higher wages attract more able applicants as measured by their IQ, personality, and proclivity toward public sector work—i.e., we find no evidence of adverse selection effects on motivation; higher wage offers also increased acceptance

rates, implying a labor supply elasticity of around 2 and some degree of monopsony power. Distance and worse municipal characteristics strongly decrease acceptance rates but higher wages help bridge the recruitment gap in worse municipalities.

I am not willing to argue that paying U.S. government officials more would necessarily improve our political system. But, just as it seems a bad idea to pay a schoolteacher less than a commensurately talented person can make in other fields, it is probably a bad idea to expect that enough good politicians and civil servants will fill those jobs even though they can make a lot more money doing something else.

There's an even more radical idea I've been thinking about for a while: What if we incentivized politicians with big cash payouts if the work they do in office actually turns out to be good for society?

One big problem with politics is that politicians' incentives are generally not aligned well with the incentives of the electorate. Voters want politicians to help solve hard problems that have long-term time frames: transportation, health care, education, economic development, geopolitical affairs, and so on. The politicians, meanwhile, have strong incentives to act in their own interests (getting elected, raising money, consolidating power, etc.), most of which have short-term payouts. So as much as we may dislike how many politicians act, they're simply responding to the incentives the system puts before them.

But what if, instead of paying politicians a flat rate for their work, thereby encouraging them to exploit their office for personal gains that may go against the collective good, we incentivized them to work hard for the collective good?

How would I go about doing this? By offering politicians the equivalent of stock options in the legislation they produce. If an elected or appointed official works for years on a project that yields good outcomes in public health or education or transportation, let's write them a big check five or ten years down the road, once those outcomes have been verified. What would you rather do: pay a U.S. secretary of education the standard $200,000 salary whether or not he does anything worthwhile—or write him a check for $5 million in ten years if his efforts actually manage to raise U.S. test scores by 10 percent?

I have run this idea by a number of elected politicians. They do not think it is entirely crazy, or at least they are polite enough to pretend they don't. I recently had the chance to talk through the idea with Senator John McCain. He listened carefully—nodding, smiling, the whole bit. I couldn't believe how engaged he was. This only encouraged me to go on and on, in great detail. Finally, he reached to shake my hand. "That's a neat idea, Steve," he said, "and good luck to hell with that!"

He turned and walked away, still smiling. I have never felt so good about being so fully rejected. I guess *that's* what it takes to be a great politician.

Chapter 2

Limberhand the Masturbator and the Perils of Wayne

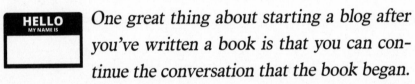 One great thing about starting a blog after you've written a book is that you can continue the conversation that the book began. A book, once it's published, is pretty much set in stone. But the blog can be updated every day, every hour. Even better: you now have an army of book readers scouring the universe for stories that confirm (or refute) what you wrote in the book. Such was the case with a *Freakonomics* chapter called "Would a Roshanda by Any Other Name Smell as Sweet?" in which we explored the impact that a person's name has on his or her life outcomes. No reader was more diligent in the pursuit of this idea than the woman who inspired the first post in this chapter.

THE NEXT TIME YOUR DAUGHTER BRINGS HOME A NEW BOYFRIEND, BE SURE TO ASK HIS MIDDLE NAME
(SDL)

I got an interesting a package in the mail recently. It came from a Texas woman named M. R. Stewart, who says she is a proud mother and a grandmother to four pit bulls.

Ms. Stewart has an unusual hobby: clipping newspaper articles of a particular ilk. She sent me photocopies of her most recent finds, all from her local newspaper, over the past few years. The articles had two things in common:

1. They were all reports of alleged crime.

2. In each case, the alleged perpetrator's middle name was Wayne.

I have to say I was stunned by the number of examples; in order to protect the potentially innocent, I will obscure their last names:

ERIC WAYNE XXXXXX: sex charges

NATHAN WAYNE XXXXXX: kidnapping and beating, homicide

RONALD WAYNE XXXXXX: triple homicide

DAVID WAYNE XXXXXX: ten years for practicing nursing without a license

LARRY WAYNE XXXXXX: homicide

PAUL WAYNE XXXXXX: theft

MICHAEL WAYNE XXXXXX: theft

JEREMY WAYNE XXXXXX: homicide

GARRY WAYNE XXXXXX: knowingly having unprotected sex when HIV positive

BRUCE WAYNE XXXXXX: homicide

JOSHUA WAYNE XXXXXX: assault of officer

BILLY WAYNE XXXXXX: homicide

BILLY WAYNE XXXXXX: assault

BILLY WAYNE XXXXXX: attempted murder and robbery

KENNETH WAYNE XXXXXX: sex assault

JERRY WAYNE XXXXXX: attempted homicide

TONY WAYNE XXXXXX: aggravated assault of grandmother in front of her grandchildren, robbery

LARRY WAYNE XXXXXX: home invasion

RICHARD WAYNE XXXXXX: police standoff

CHARLES WAYNE XXXXXX: homicide

Maybe you could assemble a list this impressive for some other middle name, but I doubt it. And of course anyone with the middle name of Wayne has a scary role model in the notorious Chicago serial killer John Wayne Gacy Jr.

Ms. Stewart also collects clippings with middle names that rhyme with Wayne: there were four DeWaynes, four Duanes, and two Dwaynes.

After going through the package, I pulled my two oldest daughters aside (they are six) and told them they were not allowed to ever have a boyfriend with the middle name Wayne. Olivia, who is obsessed with a boy named Thomas in her class, is going to check on his middle name tomorrow.

YOURHIGHNESS MORGAN
(SJD)

Thanks to our *Freakonomics* section about unusual first names—like Temptress, Shithead (pronounced shuh-TEED), and Lemonjello and Orangejello— we regularly get e-mails from readers with similar examples.

I don't think there's been a better submission than this one, courtesy of David Tinker of Pittsburgh. He sent an *Orlando Sentinel* article about a sixteen-year-old student athlete in Bushnell, Florida, named Yourhighness Morgan. He has a younger brother named Handsome, and cousins named Prince and Gorgeous.

(FWIW, I grew up as a farm kid and we had a pig named Handsome.)

Yourhighness often goes by YH for short, and also sometimes Hiney—which, to the friends and family who call him this, apparently doesn't mean "tush" or "derriere," which it did in my house.

I like Yourhighness so much that I am going to try to get my kids to call me that for a while.

In other strange-name news, there's a sad *San Diego Tribune* article (sent to us by one James Werner of Charlottesville, Virginia) about a gang murder. The victim's name was Dom Perignon Champagne; his mother's name is Perfect Engelberger.

WHAT A HEAVENLY NAME
(SJD)

What child hasn't played around with the spelling of his or her name—wondering, for instance, how it would sound if it were spelled backward? (I admit that I signed some school papers "Evets Renbud" when I was a kid.) Well, now it seems that at least 4,457 parents last year did the work for their children, giving them the name "Nevaeh," which is "Heaven" spelled backward. Jennifer 8. Lee (who is herself nomenclaturally blessed) has the story in *The New York Times*, showing an absolutely remarkable spike in popularity in a new name—from 8 instances in 1999 to 4,457 in 2005.

> "Of the last couple of generations, Nevaeh is certainly the most remarkable phenomenon in baby names," said Cleveland Kent Evans, president of the American Name Society and a professor of psychology at Bellevue University in Nebraska. . . . The surge of Nevaeh can be traced to a single event: the appearance of a Christian rock star, Sonny Sandoval of P.O.D., on MTV in 2000 with his baby daughter, Nevaeh. "Heaven spelled backwards," he said.

The only squirrelly point in Lee's article is the assertion that "Nevaeh," the seventieth ranked name for U.S. baby girls, is now more popular than "Sara"—which is true, but a little misleading: the more common spelling of "Sarah" is still ranked No. 15.

THE UNPREDICTABILITY OF BABY NAMES (SJD)

Is it possible to predict which names will become more popular over time, and which will fall? We did make a run at predicting some of the boy and girl names that might become popular in ten years' time, based on the observation that the masses tend to choose names that first become popular among high-education, high-income parents. But trends, including naming trends, tend to march to a drummer that isn't always audible.

But if you had to pick one name in the past couple of years that you were sure would be abandoned, you might pick Katrina. Who on earth would name their baby after a hurricane that nearly wiped out an entire city?

And indeed, the name did slump in the twelve months following Hurricane Katrina, with only 850 incidences in the U.S. It slipped on the list of girls' names from number 247 to number 382. That's a pretty big drop—but why wasn't the drop even steeper?

You might think it's because parents far from the affected areas weren't all that tuned in to the hurricane and its destruction. If so, you would be wrong.

In the two states most severely affected by Hurricane Katrina, the name actually received *more* action in the twelve months following the storm than in the twelve months previous. In Louisiana, the name increased from eight incidences to fifteen, while in Mississippi, it spiked from seven to twenty-four. (I am guessing that the *rate* of Katrina namings increased even more, since lots of displaced people from both states were having babies—maybe named Katrina—elsewhere.)

Maybe new parents in Louisiana named their babies Katrina as affirmation that they'd lived through the storm, a kind of hair-of-the-dog naming treatment. Maybe they named their girls Katrina to commemorate friends or relatives who died or lost their homes. But one thing's for sure: I don't know of anyone who would have predicted that there would be more Katrinas in Louisiana and Mississippi after the hurricane. Which says at least as much about our incessant desire to predict the future as it does about the people who had babies last year.

BEAT THIS APTONYM
(SJD)

An aptonym is a name that also describes what you do. In the old days, aptonyms weren't coincidences; they were professional labels. That's why there are still so many people named Tanner, Taylor, etc. But in our culture, they are quite rare.

Which is why I got so excited yesterday when I spotted a fantastic aptonym. Flipping through the latest issue of *Good* magazine, I stopped to look at the masthead. There are two people listed under "Research," which usually means fact-checking in magazine talk. One of the names is . . . Paige Worthy. That is: if a fact doesn't get past Paige Worthy, then it's not page-worthy, at least not for *Good*.

Is this a gag name? I doubt it—all the other names on the masthead look legit—and I sincerely hope not. Can you offer a better aptonym than Paige Worthy?

At the end of this post, we announced a contest invit-ing readers to submit the best aptonyms they'd ever come across. The submissions would be judged by a blue-ribbon panel of naming experts (a.k.a. Dubner and Levitt), and the winners would be sent a piece of Freakonomics swag.

ANNOUNCING THE WINNERS
OF OUR APTONYM CONTEST
(SJD)

We recently blogged about a fact-checker named Paige Worthy and asked you to send us your own aptonyms. You responded mightily, with nearly three hundred submissions. Judging from this sample, the dentists, proctologists, and eye doctors of America seem par-ticularly prone to aptonymous behavior. Below you will find the best submissions. But first, a little more information about the person who got this all started, Paige Worthy:

Yes, she is real, and that is her real name. Not only is she a researcher for *Good* magazine, but she is also a copy editor for *Ride* and *King* magazines, both of which are geared toward a black male readership. The first is a car magazine; the second is a lad magazine, referred to in some quarters as Blaxim. "I'm a white girl, by the way," Paige wrote in to say. She lives in New York and is originally from Kansas City—where, she says, "I worked at a little community outfit called the Sun Tribune Newspapers, where I was a copy editor and page designer, so my name was doubly apt at that point."

So, because she is real and because her name is the perfect aptonym, Paige Worthy definitely gets whatever Freakonomics prize she wants. The other winners:

Limberhand the Masturbator

A reader named Robbie wrote in to tell of an Idaho court case about expected privacy in a public restroom stall. This was in relation to the Larry Craig brouhaha. Here's a brief excerpt from the Idaho case:

> The defendant was arrested for obscene conduct after an officer observed him, through a four-inch hole in a stall partition, masturbating in a public

restroom. This Court determined that Limberhand had a legitimate expectation of privacy in the restroom stall notwithstanding the existence of the hole.

That's right: the man in the stall who was arrested for masturbating in public was named Limberhand.

(Honorable mention in the Below-the-Belt category goes to the reader who wrote this: "I once edited a medical journal article about penile lengthening, written by Dr. Bob Stubbs. Best of all, he learned his technique from a Chinese plastic surgeon, Dr. Long.")

Eikenberry the Funeral Director

A reader named Paul A. wrote this: "In Peru, Indiana, there's a funeral home director whose last name is 'Eikenberry' (pronounced 'I can bury'). He's actually part of a partnership, and the funeral home is called (drumroll, please) 'Eikenberry Eddy.'"

(Honorable mention in the Six-Feet-Under category to the reader who writes: "In my hometown [Amarillo, TX], there is a funeral director called Boxwell Brothers. This one can't be beat.")

Justin Case the Insurance Guy

I'm not sure this one is real but I will assume that Kyle S., the reader who sent it in, is an honest man: "My

State Farm agent's name is Justin Case . . ." Enough said.

And finally, though I said we'd give just three prizes, there were so many aptonymous dentists that I think we have to stretch the winners to four. Here's my favorite:

Chip Silvertooth

A reader named Scott Moonen writes: "My former dentist was named Eugene Silvertooth. From childhood he had the nickname Chip Silvertooth."

(The honorable mention for dentists goes to a reader named Anshuman: "Unfortunately, I moved away from San Francisco and had to leave my dentist, Dr. Les Plack. He was born for the job, right?")

Chapter 3
Hurray for High Gas Prices!

If there is one thing economists think they know everything about, it is prices. To an economist, there is a price on everything, and everything has a price. Regular people think of prices as what you grumble about at the store; economists think of prices as the logic that organizes our world. So of course we've had a lot to say on this topic over the years.

SOMEBODY HATES ME $5 WORTH
(SDL)

There is a website—one that is so stupid I feel embarrassed to even give it free publicity—called www. WhoToHate.com. The idea behind the site is you pay them five bucks, write in the name of someone you hate, and the website writes to the person telling them that there is someone who hates them.

I got one of those hate mails today, meaning that someone hates me enough to be willing to pay five dollars to have me receive such an e-mail.

From an economic perspective, it is an interesting product they are providing. Does the person spending the five dollars get utility from the act of declaring (albeit completely anonymously) their hatred? Or does the utility come from the (real or imagined) pain on the part of the recipient when he or she discovers the depth of another's hatred?

For someone who actively hates me, the only source of satisfaction would be the first channel. I already get loads of hatred coming my way every day—hatred that is far more vicious than this whimsical e-mail I received after they spent five dollars. Indeed, the fact that the person who hates me identified me as Steve Levitt from California (where I lived only briefly while visiting Stanford a number of years ago) actually gave me a good chuckle.

It got me thinking. Maybe the website would be well served to allow the hater to make a payment greater than five dollars. By paying fifty dollars to demonstrate their hatred, and relaying that information to the hated person, it could really send the message. Perhaps, though, haters prefer to send ten separate five-dollar messages to create the impression that everyone hates you a little, rather than one person hates you a lot.

What saddens me about the website is that where it could really have some bite is for some innocent teenager who is singled out for hatred by his or her peers. For a person who only gets a few e-mails a day to begin with, receiving ten or twelve e-mails saying that anonymous people hate you might be pretty discouraging.

The good news is that apparently not many people feel enough hatred to want to spend five dollars to make that hatred known. The current list of the ten most hated people includes some well-known names (I've omitted the people I have never heard of, out of fear they are the innocent teens I mentioned). Here is the list, with their number of hatreds:

George Bush   (7)
Hillary Clinton   (3)
Oprah Winfrey   (3)
Gloria Steinem   (3)
Barbara Boxer   (2)

So even with all the people who hate George Bush, only seven people have been willing to pony up the five dollars! To make the top-ten list, you only need two people to hate you. That shouldn't be hard for me. I'm already halfway there.

IF CRACK DEALERS TOOK LESSONS FROM WALGREENS, THEY REALLY WOULD BE RICH
(SJD)

Not long ago, I was chatting with a physician in Houston, the sort of older gentleman family doctor you don't see much of anymore. His name is Cyril Wolf. He's originally from South Africa, but other than that, he struck me as the quintessential American general practitioner of decades past.

I'd asked him a variety of questions—what's changed in recent years in his practice, how managed care has affected him, etc.—when suddenly his eyes fired up, his jaw set tight, and his voice took on a tone of great exasperation. He began to describe a simple but huge problem in his practice: a lot of generic medications are still too expensive for his patients to afford. Many of his patients, he explained, must pay for their drugs out of pocket, and yet even the generic drugs at pharmacy chains like Walgreens, Eckerd, and CVS could cost them dearly.

So Wolf began snooping around and found that two chains, Costco and Sam's Club, sold generics at prices far, far below the other chains. Even once you factor in the cost of buying a membership at Costco and Sam's Club, the price differences were astounding. (Nor, apparently, do you need to be a member of either store to use their pharmacy, although membership does bring a further discount.) Here are the prices Wolf found at Houston stores for ninety tablets of generic Prozac:

Walgreens: $117
Eckerd: $115
CVS: $115
Sam's Club: $15
Costco: $12

Those aren't typos. Walgreens charges $117 for a bottle of the same pills for which Costco charges $12.

I was skeptical at first. Why on earth, I asked Wolf, would anyone pay $100 extra—probably every month— to fill a prescription at Walgreens instead of Costco?

His answer: if a retiree is used to filling his prescriptions at Walgreens, that's where he fills his prescriptions, and he assumes that the price of a generic drug (or, perhaps, any drug) is pretty much the same at any pharmacy. Talk about information asymmetry; talk about price discrimination!

I had meant to write about this, and had collected a few relevant links: a TV news report in Houston about Wolf's discovery; an extensive price comparison compiled by a TV news reporter in Detroit; a *Consumer Reports* survey; and a research report on the subject from Senator Dianne Feinstein.

But I had forgotten all about this issue until I read a comprehensive *Wall Street Journal* article that does a good job of measuring the difference in prices between chains. Most of the differences aren't as drastic as Wolf's example, but are often still huge. Perhaps the most interesting sentence is this one:

> After a call from a reporter, CVS said it would drop its simvastatin price [from $108.99] to $79.99, as part of an "ongoing price analysis."

So that's what it's called: "ongoing price analysis." I'll have to remember that the next time my kids catch me trying to buy a two-dollar toy when I'd promised one for twenty.

THE NEW-CAR MATING DANCE
(SDL)

My car is ten years old, so I went out to buy a new one this weekend. In *Freakonomics* and *SuperFreakonomics*, we write a lot about how the Internet has changed markets in which there are information asymmetries. Buying a new car gave me the chance to see firsthand these forces at work in the new car market.

I was not disappointed. I already knew what kind of car I wanted. Within fifteen minutes and at no cost, using sites like TrueCar and Edmunds, I not only had a good idea of a fair price to pay for the car, but also was able to notify some local car dealerships that I was interested in quotes.

Just a few minutes later, one car dealership offered to sell me the car at $1,300 under invoice. That seemed like a good place to start, but before I could even round up the kids to drag them to the dealership, another dealership called, and when they heard the offer from the first dealership, they beat that offer by a few hundred dollars. I called the first dealership back and got voice mail, so we headed off to the second one. I figured I was still far away from a final price, but I was off to a good start without even leaving our house.

I learned a lot about buying cars the last time I bought one—the various lies that dealerships tell with respect to invoice prices; the ridiculous game of cat and mouse with the salesperson trotting off to talk to the manager, etc. I abhorred the process the last time we needed a car, but this time, thinking about it more intellectually, I was eager to take part in the elaborate ritual associated with buying a new car.

Perhaps my willingness to haggle stemmed from my unlikely triumph the last time around. I had gotten an estimate faxed to me—this was pre-Internet—of what a fair price was to pay for the car. Stupidly, I left the sheet of paper at home, but I thought I remembered the price. I fought hard for that price: threatening repeatedly to leave, back and forth and back and forth, and finally I got the dealer within a few hundred dollars of the price I remembered. When I got home, I discovered that my memory had transposed two digits; the price on the fax was actually $2,000 higher than the one I managed to bargain. Mistakenly believing a fair price was $2,000 lower than it really was, I had bargained incredibly hard and gotten a ridiculously good deal. Leaving that fax at home was worth thousands of dollars.

So I got to the car dealership and sat down to bargain. The salesman explained to me that the price they were offering was well below invoice, discreetly showing me some pricing documents stamped "confidential," emphasizing how much money they were going to lose on the car. I replied that he knew as well as I did that the invoice price he was quoting me was not what the dealership paid for the car. I asked him to simply give me his best offer. He disappeared for a while, ostensibly to talk to his boss, but probably to catch up on who was winning the baseball game.

He was gone just long enough for a third dealership to e-mail me a price quote. This one was $1,500 lower than the current best offer from the dealership where I was sitting. He came back and said the best he could do would be to go $200 lower. I said, "That's not going to work because another dealer just offered to beat that price by over $1,000." I handed over my phone to show him the e-mail. He disparaged the other dealership for a while, and then he went and found the boss. The boss assured me that their last offer was the very best they could do—it was a generous offer for a dozen different reasons.

I said, "That's all fine, but if you don't do better, I'm going to walk out of here and go to the other dealership."

By my estimation, we were now about halfway through the mating ritual. In about fifteen minutes, after a lot of huffing and puffing, we would get the car for the price the third dealership had offered. That would still probably be paying way too much, but I was willing to accept that outcome.

"So I'm going to walk out," I reiterated.

"Okay," the manager said. "If it doesn't work out at that other dealership, come back and we will sell you the car at the price we offered before."

I stood up and began to gather the kids, all part of the tough bargaining act. They simply watched me, as if they had forgotten that this was all part of the ritual. Even if they had forgotten their lines, I still remembered mine. "We both know that if I walk out of here today, I am never coming back."

To which the guy simply said, "We're willing to take our chances on that."

And I walked out.

I was shocked. This dealership had sent me a price over the Internet, come down only $200, and then smiled as I left to buy a car from another place. Given that, I figured the new dealer must be giving us a really good deal. I didn't have the energy to start another mating dance with the new dealership, so I simply accepted their offer without bargaining. I pick the car up on Tuesday.

FOR $25 MILLION, NO WAY—BUT FOR $50 MILLION I'LL THINK ABOUT IT
(SDL)

At least for me, there are not too many questions that would lead me to respond, "For twenty-five million, no way—but for fifty million I'll think about it." Twenty-five million dollars is so much money that it's hard to think about what you would do with it. It sure would be nice to have the first $25 million but I'm not sure why I'd need the second $25 million.

The U.S. Senate is hoping there are some folks in Afghanistan or Pakistan who don't see it that way. Frustrated by the failure of the $25 million bounty on Osama bin Laden to lead to his capture, the Senate has voted 87–1 to raise the bounty to $50 million. (The lone dissenter was Jim Bunning, a Republican from Kentucky.)

At one level, you have to applaud this move by the government. To a Pakistani peasant, $50 million is an unthinkably large amount of money. To the U.S. government, which is spending $10 billion a month in Iraq, $50 million is next to nothing. If one of the major goals of the Iraq war was to get rid of Saddam Hussein, think how much cheaper it would have been to offer a reward of, say, $100 billion to anyone who could get him out of office by whatever means they saw fit. Saddam himself might have graciously accepted the offer and traded the hassles of running a country for a pleasant $100 billion pension and a well-appointed French manor.

Indeed, we have written before about the virtues of offering big prizes to encourage people to work on problems, whether it is curing disease or improving Netflix's algorithms.

On the other hand, if I can't tell the difference between $25 million and $50 million, I can't imagine that upping the ante will push a wavering Pakistani over the edge of collaborating with the U.S. government.

Much more important, but harder to do, would be to find a way to make it credible that we will actually pay the bounty. I'm sure there is plenty of discretion in deciding to whom and how much of that bounty gets paid. For instance, if I did some statistical analysis that somehow narrowed down his whereabouts to within one thousand yards, and then the Navy SEALs canvassed that area and found him, would I get the money? I'm not so sure they would give it to me. I'm guessing the Pakistani peasant who has some information on Bin Laden probably shares my doubts.

Indeed, no bounty was ultimately paid. As reported by ABC News, "the raid that killed the al Qaeda leader in Pakistan on May 2 [2011] was the result of electronic intelligence, not human informants . . . The CIA and the military never had an al Qaeda operative as an informer willing to give him up."

HOW MUCH WOULD PEPSI PAY TO GET COKE'S SECRET FORMULA?
(SDL)

Some dastardly Coca-Cola employees recently got nabbed trying to sell corporate secrets to Pepsi. Pepsi turned the bad guys in and cooperated in the sting operation.

Did the executives at Pepsi give up the chance to make huge profits at Coke's expense in order to "do the right thing"?

I had lunch yesterday with my friend and colleague Kevin Murphy yesterday. He made an interesting point: knowing Coke's secret formula is probably worth almost nothing to Pepsi. Here is the logic.

Let's say that Pepsi knew Coke's secret formula and could publish it so that anyone could make a drink that tasted just like Coke. That would be a lot like what happens to prescription drugs when they go off patent and generic drug companies come in. The impact would be that the price of real Coke would fall a lot (probably not all the way to the price of the generic Coke knock-offs). This would clearly be terrible for Coke. It would probably also be bad for Pepsi. With Coke now much cheaper, people would switch from Pepsi to Coke. Pepsi profits would likely fall.

So if Pepsi had Coke's secret formula, they wouldn't want to give it away to everyone. What if they instead kept it to themselves and made their own drink that tasted exactly like Coke? If they could really convince people that their drink was identical to Coke, then the new Pepsi-made version of Coke and the Real Thing would be what economists call "perfect substitutes." When two goods are essentially interchangeable in consumers' minds, that tends to lead to fierce price competition and very low profits. Neither Coke nor the Pepsi knockoff of it would be very profitable as a consequence. With the price of Coke lower, consumers would switch away from the original Pepsi to either Coke or the new Pepsi-made Coke knockoff, which would be far less profitable than original Pepsi anyway.

In the end, both Coke *and* Pepsi would likely be worse off if Pepsi had Coke's secret formula and acted on it.

So, maybe the executives at Pepsi were acting morally and honorably when they turned in those suspected of stealing Coke's secrets.

Or maybe they are just good economists.

CAN WE PLEASE GET RID OF THE PENNY ALREADY?
(SJD)

What began as a casual observation somehow turned into a crusade, with Dubner becoming an unofficial spokesman for the abolition of the penny. During a 60 Minutes segment on the subject, he said the U.S. suffers from "pennycitis," and that the penny is about as useful as "having a fifth and a half finger on your hand." Below are excerpts of various anti-penny posts.

Whenever I get change for a dollar, I ask the cashier to keep the pennies. They aren't worth my time, or hers, or yours. Sometimes the cashier refuses for bookkeeping purposes, in which case I politely accept the pennies and then throw them in the nearest trash can. (Is this illegal? If so, then I guess we should start arresting people for throwing money in wishing wells, too.)

If I were the type of person who regularly a) loaded up my pocket every day with loose change or b) brought all my loose change to a bank or supermarket coin machine, then it might be worthwhile to keep the pennies. But I'm not, and so it's not. These facts, coupled with the reality of inflation, have led me to wish for years that the penny would be abolished, and probably the nickel, too. (When we were kids, playing Monopoly, we never used the one-dollar bills; did you?)

There are all kinds of reasons to get rid of the penny, but perhaps the only one you need to know is that it costs the U.S. Government a lot more than one cent to make a penny. Considering that we lose money every time a penny is made, and that they aren't useful in any meaningful way, it seems like a no-brainer that we should get rid of the penny. Inflation has rendered it a bad idea, for both producer and consumers.

But I am happy to see that there is a sensible alternative to throwing away loose change: "rebasing" the penny to make it worth five cents. The plan comes courtesy of François Velde, an economist at the Chicago Fed. I'd like to think that the serious people in charge of our nation's currency will take this argument seriously, but considering what I know about the penny, and politics, and inertia, I'm not holding my breath.

Why does the U.S. still use pennies? One big reason: lobbyists. I recently appeared on a *60 Minutes* segment called "Making Cents." I discussed the foolishness of keeping the penny; but *60 Minutes* included the pro-penny position as well. Here's an excerpt:

> Mark Weller is the voice of "Americans for Common Cents," a pro-penny group that claims that rounding up will cost Americans $600 million a year . . . He says without the penny, charities, too, would suffer, on the theory that people are less likely to donate as many nickels. As it is, penny drives around the country collect tens of millions of dollars a year for medical research, for the homeless, for education . . .
>
> But as Weller freely admits, he's got a financial interest in the high cost of penny pinching: Weller

is a lobbyist for Jarden Zinc, the Tennessee company that sells those little blank discs for the mint to turn into Lincoln pennies.

I guess instead of wasting my time arguing against the penny, I should just buy some zinc futures.

The Great Penny Debate continues to limp along. One hundred million pennies, collected by schoolchildren, were put on display at Rockefeller Center. Meanwhile, lots of people continue to argue for elimination of the penny.

I am firmly on the abolitionists' side. The only reasons I can think of for keeping the penny are inertia and nostalgia. Talk about deadweight loss!

The most ridiculous pro-penny defense I've seen in a while appeared in a recent full-page ad in the *Times*. It was taken out by Virgin Mobile, which was promoting its texting service as so cheap to use that it made even a penny worth keeping. The headline read:

**"New Legislation Will Attempt to
DO AWAY WITH THE PENNY.
What's Next, Puppies and Rainbows Too?"**

Here is the line that caught my attention:

And what does America think? 66%* of our population wants to keep the penny and 79% would stop to pick one up off the ground.

If you follow that asterisk to the bottom of the ad, here's what you find:

*Source: The 8th Annual Coinstar National Currency Poll

For those of you who don't know, Coinstar is the company that puts change machines in supermarkets, in which you can dump your coin jar and receive a receipt that you take to the cash register for folding money. Coinstar apparently takes a commission of 8.9 percent for providing this service.

While the Coinstar National Currency Poll is said to be compiled by an independent market research organization, I am somehow not very surprised that a survey commissioned by a company that makes money from coin harvesting is able to produce a result saying that two-thirds of Americans "want to keep the penny."

I never set out to be anti-penny, but somehow it happened, and I now publicly rant whenever possible that the penny should be eliminated.

While I stand by my belief that the penny is lousy as currency, someone has finally come up with a use for pennies that has made me reconsider my extinction argument: make a floor out of them!

The penny floor can be found at the Standard Grill at the new Standard Hotel in New York, the one straddling the High Line. The Standard tells us that it used 250 pennies per square foot, or 480,000 pennies in all.

For those of you thinking about a home renovation, that's $2.50 per square foot in flooring materials. That stacks up pretty well against glass tile ($25), polished marble ($12), porcelain ($4), or even prefinished walnut ($5). It tells you something about the penny's uselessness as a currency that, even though it is actual money, it is still cheaper than all these other materials to make a floor out of.

PLANNED PARENTHOOD GETS FREAKY!
(SDL)

For a long time, the pro-life movement has had a keen sense of how people respond to incentives. Protesters outside of clinics proved to be a very effective strategy for raising the social and moral costs of seeking an abortion.

Now a Planned Parenthood clinic in Philadelphia has come up with a very clever strategy for fighting back, called Pledge-a-Picket. As they explain:

Every time protesters gather outside of our Locust Street health center, our patients face verbal attacks from them. They see graphic signs meant to confuse and intimidate . . . We are all called murderers, are lectured to about committing sins, and are told we will pay the "ultimate price" for our actions.

Here's how it works: You decide on the amount you would like to pledge for each protester (minimum 10 cents). When protesters show up on our sidewalks, Planned Parenthood Southeastern Pennsylvania will count and record their number each day . . . We will place a sign outside the health center that tracks pledges and makes protesters fully aware that their actions are benefiting PPSP. At the end of the two-month campaign, we will send you an update on protest activities and a pledge reminder.

My prediction: abortion clinics around the country will soon be adopting this approach. What I think is so clever about this approach is the way it transforms the outrage, anger, and helplessness that ardent pro-choice folks feel toward the protesters into a financial incentive that works on behalf of the pro-choice people and against the protestors. On the margin, I think that donations will be higher because potential donors can then derive pleasure from the presence of the protesters, or at least less pain. It is empowering. On the other hand, if I am a protester, I will hate the idea that what I am doing may be making Planned Parenthood stronger, decreasing the utility of the protest.

LOST: $720 BILLION. IF FOUND, PLEASE RETURN TO OWNER, PREFERABLY IN CASH
(SDL)

According to the S&P/Case-Shiller index of housing prices, home prices fell by about 6 percent in the U.S over the course of 2007. By my rough calculations, that means that homeowners have lost about $720 billion in wealth as a consequence. That is about $2,400 for every person in America, and $18,000 for the average homeowner.

Relative to stock market declines, however, that loss of $720 billion in a year doesn't look quite so big. The total market capitalization of U.S. stock markets is the same order of magnitude as the total value of the housing market (between $10 and $20 trillion). In one week during October 1987, the U.S. stock market lost over 30 percent of its value.

The $720 billion figure is also about the same magnitude as the amount of money the U.S. government spent on the war in Iraq for the first few years.

If you are a homeowner, how bad do you feel about this? You should feel pretty bad, but I'm guessing you would feel a lot worse in the following scenario: home prices did not fall at all last year, but one day you took $18,000 out of the bank to pay cash for a new car, and someone then stole your wallet with the $18,000 in it. At the end of the day, your wealth would be the same (down $18,000, either from depreciation of the value of your home or because the money was stolen), but one loss is psychologically far worse than the other.

There are many possible reasons for why it doesn't hurt so much to lose money on an asset like a house. First, it isn't very tangible, since no one really knows what their house is worth anyway. Second, it hurts less whenever everyone else is also losing on their houses. (I once heard a very rich person say that he didn't care about his absolute wealth, only what his ranking was on the *Forbes* list of richest people.) Third, you can't really blame yourself for house prices falling, but you could second-guess your decision to carry around $18,000 in cash. Fourth, the fact that a thief has your money might make it worse than the money just evaporating into space, like it does when house prices fall. There are probably other reasons as well.

More generally, the economist Richard Thaler coined the phrase *mental accounts* to describe the way in which people seem to treat different assets as non-fungible, even though in principle it seems like they should be. Although my economist friends make fun of me for it, I definitely use mental accounts myself. For me, a dollar made playing poker means much more than a dollar earned from the stock market going up. (And a dollar lost playing poker is likewise far more painful.)

Even people who deny that they are affected by mental accounts often fall prey to them. I've got a buddy in that category who won a big bet on NFL football (big relative to his usual football bet, but very, very small relative to his overall wealth) and the next day he spent the proceeds on a fancy new driver.

What does this all mean for housing prices? Well, if prices start going back up, it would be a lot more fun if the price increases came in the form of little packets of cash dropped outside your front door with the morning newspaper, rather than via house appreciation. I suppose all those people who took out home-equity loans figured this out a long time ago.

HOW IS A CANADIAN ART-POP SINGER LIKE A BAGEL SALESMAN?
(SJD)

Much like Paul Feldman, the economist-turned-bagel salesman we wrote about in *Freakonomics,* the singer-songwriter Jane Siberry has decided to offer her wares to the public via an honor-system payment scheme. She gives her fans four choices:

1. free (gift from Jane)

2. self-determined (pay now)

3. self-determined (pay later so you are truly educated in your decision)

4. standard (today's going rate is about .99)

Then, cleverly, she posts statistics on payment rates to date:

% Accepting gift from Jane: 17%
% Paid by determining price: 37%
% Paying Later: 46%

Avg. price per track: $1.14
% paid below suggested: 8%
% paid at suggested: 79%
% paid above suggested: 14%

Even more cleverly, Siberry posts the average payment rate for each song as you pull your payment option from the drop-down menu—another reminder that, hey, you're more than welcome to steal this music but here's how other people have acted in the recent past.

It seems that Ms. Siberry grasps the power of incentives quite well. This allows for at least a couple of interesting things to happen: people can decide what to pay after they hear the music, and see how much it's worth to them (it looks like people generally pay the most per song under this option); and it takes the variable-pricing scheme that economists love and puts it in the hands of the consumer, not the seller.

I think record companies will need a lot more convincing before they're willing to try this model on a large scale. Presumably, Jane Siberry fans who go to her website to get her music are a deeply self-selecting lot, far more devoted than the average downloader. But as desperate as the record companies are, I wouldn't be surprised to see more of this in the future.

TWO DAYS LATER . . .

JANE SIBERRY SNAPS
(SJD)

Apparently, Jane Siberry doesn't appreciate people calling attention to her website, which allows people to pay as they wish to download Siberry's music. I liked the idea, and blogged about it. But here's what Siberry wrote on her MySpace journal today:

> The "self-determined pricing" policy of the store is in the spotlight again, freakonomics has an online article; abc news emailed. I don't want the attention. I think I'll change the pricing to "you can pay me all you want but i'm not going to let you hear it."

Youch. Regrets, Ms. Siberry. Seems like we have a lousy track record with pop singers—anybody remember when Levitt announced that Thomas Dolby was releasing a new record, an announcement that turned out to be 100 percent wrong?

I guess we should give up pop singers and stick to crack dealers, real-estate agents, and poker cheats.

HOW MUCH TAX ARE ATHLETES WILLING TO PAY?
(SJD)

The Laffer curve is a unicorn-y concept that seeks to explain the rate of taxation at which revenues will fall because earners either move away or decide to earn less (or cheat more, I guess).

If I were a tax scholar interested in this concept, I would be taking a good, hard look at the current behavior of top-tier professional athletes. Boxing is particularly interesting because it allows a participant to choose where he performs. If you are a pro golfer or tennis player, you might be inclined to skip a particular event because of a tax situation, but you generally need to play where the event is happening. A top-ranked boxer, meanwhile, can fight where he gets the best deal.

Which is why it's interesting to read that Manny Pacquiao will probably never fight in New York—primarily, says promoter Bob Arum, because of the taxes he'd have to pay. From the *Wall Street Journal*:

> Manny Pacquiao has won fights in California, Tennessee, Texas and Nevada, not to mention Japan and his native Philippines. But with Pacquiao in New York this week to promote his next fight—a November bout in Macau against Brandon Rios—Pacquiao's team said Barclays Center and the Garden were two venues where he wouldn't fight because he would have to pay the state's tax rate in addition to federal taxes. "He'd have to be a lunatic," said Bob Arum, Pacquiao's promoter.

In an *L.A. Times* article, Arum says that Pacquiao may never fight anywhere in the U.S. again:

"By fighting outside the country, as he's doing in this Rios fight, Manny doesn't have to pay U.S. taxes anymore—at a rate of 40% for a foreign athlete.

"If this pay-per-view and other things take off like we think they may, I can't imagine Pacquiao will ever again fight in the U.S."

There are of course other factors at play besides taxes—gambling, for one, which is a big reason that Macau has become such a boxing center. But whatever you think of the Laffer curve, it's hard to ignore the variance in tax rates around the world, especially for athletes who might earn a lot of money in a short time.

In January, the golfer Phil Mickelson said he was "going to have to make some drastic changes" to deal with federal and California tax hikes (he lives in California). "If you add up all the Federal and you look at the disability and the unemployment and the Social Security and the state, my tax rate's 62, 63 percent," he said.

His accounting was challenged and Mickelson, one of the most popular golfers ever, was widely spanked for publicly airing his tax dissent. So last month, when he won back-to-back tournaments in Scotland (the Scottish Open and the Open Championship), he kept quiet. But the media did the speaking for him. In *Forbes*, Kurt Badenhausen wrote a (very good) article about Mickelson's British tax tab, estimating that he'd pay, in total, about 61 percent tax on his nearly $2.2 million in earnings. And Badenhausen identifies this interesting wrinkle:

> But that's not all. The U.K. will tax a portion of his endorsement income for the two weeks he was in Scotland. It will also tax any bonuses he receives for winning these tournaments as well as a portion of the ranking bonuses he will receive at the end of the year, all at 45% . . .
>
> The U.K. is one of few countries that collects taxes on endorsement income for non-resident athletes that compete in Britain (the U.S. also does). The rule has kept track star Usain Bolt from competing in Great Britain since 2009, outside of the 2012 Summer Olympics when the tax was suspended as a condition for hosting the Games. Spain's Rafael Nadal has also allowed U.K. tax policy to dictate his tennis playing schedule.

And let's not forget that the greatest endurance athlete of our era, Mick Jagger, fled the U.K. years ago because of tax considerations (and, also, the police there kept arresting him and his mates).

PRICING CHICKEN WINGS
(SDL)

The other day, I stopped by a local fried chicken joint, Harold's Chicken Shack. Just to give you a sense of what sort of restaurant this is, there is a layer of bulletproof glass separating the workers and the customers. They don't cook the chicken until you order, so I had five or ten minutes to kill waiting for my food.

One of the items on the menu is a chicken-wing dinner. With each dinner, you get a fixed amount of french fries and coleslaw.

The two-wing meal costs $3.03. The three-wing meal costs $4.50.

Since the only difference between the two meals is one extra wing, with that third wing costing the customer $1.47. I thought this was interesting, because if each of the first two wings were priced at $1.47 each, then the implied price of the french fries and coleslaw is a combined 9 cents. So it seems like Harold's is implicitly charging more for the third wing than for the first two wings, which is unusual since firms generally give quantity discounts.

I read further down the menu:

two-wing meal	$3.03
three-wing meal	$4.50

four-wing meal $5.40
five-wing meal $5.95

The four-and five-wing meal prices are more in line with how firms usually price.

So what do you think Harold's charges for a six-wing meal? Here's the answer:

six-wing meal $7.00

Definitely most bizarre. When economists see things that don't make any sense, we can't help but think of some story that rationalizes the seemingly odd behavior. Maybe Harold's prices the six-wing meal high because it is worried about obesity? Not likely, since every item on the menu is fried. Is the sixth wing especially big or tasty? Is demand by people who order six wings more inelastic?

Perhaps some clues could be found in the pricing of other items. Fried perch are sold in a similar fashion to fried chicken, again with french fries and coleslaw. Here is how perch is priced:

2-piece perch meal: $3.58
3-piece perch meal: $4.69
4-piece perch meal: $6.45

So you get that third piece of perch cheap, but they nail you on the fourth piece. This certainly hints at Harold's thinking there is some logic to this sort of pricing.

Ultimately, though, my guess is that the person who chose these prices was just confused. One thing I have realized as I have worked more with businesses is that they are far from the idealized profit-maximizing automatons of economic theory. Confusion is endemic to firms. After all, firms are made up of people, and if people are confused most of the time by economics, why wouldn't that carry over to firms?

WHY ARE KIWIFRUITS SO CHEAP?
(SJD)

I've been eating a lot of kiwifruits lately. (You may also know them as the Chinese gooseberry.) At the corner deli near my home on the West Side of Manhattan, I can buy three for a dollar. They are delicious. Unless the stickers are lying, they come from New Zealand. At thirty-three cents apiece, a New Zealand kiwifruit costs less than the price of mailing a letter to the East Side of Manhattan. (And believe me, I consider a first-class stamp one of the greatest bargains ever.) How on earth can it cost so little to grow, pick, pack, and ship a piece of fruit across the world?

To make fruit matters more complicated, I can buy one banana (also imported) and one kiwifruit for about the same price as one apple, which may well have been grown as near as upstate New York. So I wrote to Will Masters, a food economist at Tufts University's Friedman School of Nutrition.

Most economists, as I'm sure you know, reply to such queries in verse, and Will is no exception:

> *Damn supply and damn demand:*
> *Why cheap hogs and costly ham?*
> *Bargain wheat, expensive flour,*
> *The oldest villain's market power.*

> *Just one seller makes us nervous,*
> *Like that U.S. Postal Service:*
> *They may offer bargain prices,*
> *But who disciplines their vices?*

> *Middlemen have long been blamed*
> *For every market that's inflamed,*
> *Yet better explanations come*
> *From many a Hyde Park alum.*

> *Modern views from Chicago-Booth*
> *Give a nuanced view of truth,*
> *Steven Levitt and John List*
> *Made each of us a freakonomist.*

We let data speak its mind
No matter what Friedman opined
And find the price of fruit and veg
To be driven by the market's edge.

Like the tail that wags the dog,
Marginal thinking clears the fog:
Sellers, buyers, traders too,
Interact and prices ensue.

A kiwi costs 33 cents
Simply because no one prevents
Another farm or New York store
From entering and selling more.

In contrast apples may be dear,
For reasons that will soon be clear:
Picking them's below our station,
To lower costs we need migration.

Bananas have a different story,
Seedless magic, breeder's glory,
Cheap to harvest and to ship,
Who cares if workers get paid zip?

Each crop's method of production,
Where it grows and how it's trucked in,
Satisfies some needs quite cheaply
While other costs will rise more steeply.

A buyer's choices matter too,
For nonsense stuff like posh shampoo,
Prices are not down to earth,
The more you pay the more it's worth.

Behavior is as behavior does,
Maybe some things are "just because,"
Much of life's a mystery,
A habit due to history.

For prices, though, it's competition
Plus tariffs set by politicians,
That determines whether we see
Such delightfully cheap kiwi.

Bravo.

PETE ROSE PROVIDES A LESSON IN BASIC ECONOMICS
(SDL)

Some time ago, Pete Rose signed a bunch of baseballs with the inscription "I'm sorry I bet on baseball." According to media reports, he gave these balls to friends and never intended them to be sold for profit.

But the estate of someone who received some of these balls decided to put thirty of them up for auction.

There was speculation that they would sell for perhaps many thousands of dollars.

That is when Rose himself stepped in and delivered one of the fundamental lessons in economics: as long as close substitutes are available, prices won't get very high.

When Rose heard that these balls were being auctioned, he offered to sell balls with the same inscription for just $299 on his own website, effectively destroying the market for the auction-bound balls. True, the newly signed balls wouldn't be perfect substitutes, because a collector could still say he had one of the original thirty. For that reason, you wouldn't expect the auction price of the old balls to fall all the way to $299. Indeed, the auction was called off and the balls were sold for $1,000 apiece.

(Hat tip to John List, the only baseball-memorabilia-salesman-turned-economist I know.)

IF ONLY GOD HAD HAD CORPORATE SPONSORSHIP . . .
(SJD)

. . . in the book of Genesis, when the world is created. Can you imagine how rich He could have gotten by selling the naming rights of every animal, mineral, and vegetable?

If God was unlucky to toil in the days before corporate sponsorship, the Chicago White Sox are not so unlucky. They have just announced that for the next three seasons, their evening home games will begin at 7:11 P.M. instead of the customary 7:05 P.M. or 7:35 P.M. Why? Because 7-Eleven, the convenience store chain, is paying them $500,000 to do so.

I've lately noticed advertisements showing up in a lot of unlikely venues: stamped onto fresh eggs and printed on airplane barf bags, for instance. But I have to admit there is something particularly creative about affixing a value to time itself, especially if you can capture that value for your own benefit.

Maybe I will write more about that tomorrow™.

WHAT CAPTAIN SULLENBERGER MEANT TO SAY (BUT WAS TOO POLITE TO DO SO) (BY "CAPTAIN STEVE")

Captain Steve is a seasoned international pilot for a major U.S. carrier and a friend of Freakonomics. (Given the sensitivity of what he writes, he prefers anonymity.) This post was published on June 24, 2009, six months after the "The Miracle on the Hudson," in which Captain Chesley Sullenberger safely landed an Airbus A320-200 in the Hudson River. Both the plane's engines had failed, due to a bird strike, shortly after takeoff from LaGuardia Airport in New York.

After reading some of the excerpts of Captain Sullenberger's various speeches, especially those of a few weeks ago with the National Transportation Safety Board, I would like to add my editorial.

Captain Sullenberger has been a class act all the way. He's not been petty, pious, or egotistical. He is, however, like most of the captains I know and, more broadly, most of the pilots I know. Why? He doesn't need to be otherwise. When someone has accomplished what he and the scores of men and women like him have accomplished, why do we need to boast?

He implies that what he did while serving as the "skipper" of US Airways flight 1549 was simply his job. He is being as honest and accurate as he can be: "Please, no fanfare, no applause, just doing my job." But what he has also alluded to in some of his speeches is that it has taken years, even decades, to prepare himself for that one single "lifetime event" of guiding his jet into the safe, smooth landing on the Hudson River.

What he is not saying is this:

We, the airline pilots, are facing a losing battle in the PR department. You believe that we make huge salaries and are treated like royalty. Pure fiction. Why have we been losing this battle for such a long time? Simple. Because most of us are like "Sully"; we don't want applause or fanfare for doing what we are trained to do. However, we do realize that we should be fairly compensated for what we have achieved to get this job and what we continue to do on a daily basis to keep it. This backlash of pilot-bashing is building to a boiling point.

Regional carriers, like the Colgan Airlines flight in Buffalo [which crashed, killing all forty-nine aboard], employ the lowest-bidder pilots. No offense to them; this is not personal. It is the system that is at fault. Money and profits at all cost.

Airline history lesson 101: it used to be, up until the mid 1980s, that a young pilot would be hired on at a major carrier, become a flight engineer (FE), and then spend a few years managing the systems of the older-generation airplanes. But he or she was learning all the while. These new "pilots" sat in the FE seat and did their job, all the while observing the "pilots" doing the flying, day in and day out.

The FEs learned from the seasoned pilots about the real world of flying into the O'Hares and LaGuardias. They learned decision making, delegation, and the reality of "captain's final authority" as confirmed in the law. When they got the chance to upgrade, they became a copilot. The copilot's duty was to assist the captain in flying; but even during their time as the new copilot, they had the luxury of the FE looking over their shoulders—i.e., more learning. This three-man-crew concept, now a fond memory in the domestic markets but used predominately in international flying, was considered one more layer of protection.

But it's gone. Now domestic flying is being shifted to the regional carriers, like Colgan, American Eagle, Comair, and Mesa, to name a few. These consist of the lowest bidders and the newest pilots flying into the harshest of environments. The airline management teams would say that it works and that this is routine flying. I beg to differ.

Analogy: you are told you need a quadruple bypass. Now you search the Internet for the cheapest price you can get, and you rush to schedule the operation because there are only two dates that you can get that cheap rate.

Do any of us do that? No. What do we do? We get second opinions, we ask who is the best in town, etc. We ask: "Is there anyone who has been doing this surgery for the last twenty to twenty-five years?" We don't say, "Let me use someone who just graduated from medical school and was rushed through residency because it will be cheaper."

Why not apply the same logic that the public uses to buy an airplane ticket to this surgery scenario? Bypass surgery is routine, right? Some surgeons do two, three, or four a day. It must be easy.

To take that a step further, how many surgeons have to retake their medical boards every nine months in order to be qualified? Airline pilots do. We are subject to simulator check-rides every nine months to demonstrate knowledge, proficiency, and ability.

How many surgeons have to take a physical exam every six months by the AMA in order to work? None! Airline pilots do. Fail your medical exam and you're done! How many surgeons (or any other critical professional, including politicians) are subject to random drug and alcohol testing? None.

Flying across the North Atlantic is routine, right? It wasn't just a short few decades ago. We, the pilots, make it routine because we have skills, experience, and training like very few others.

Gifted? No, not many of us are. But dedicated and focused upon excellence, you bet! I have told my kids one thing many times since they were little children: "I don't expect perfection, I expect excellence." I expect 100 percent effort in all you do. This is the creed of every pilot I know.

Flying from Chicago's O'Hare to Denver is routine, right? We, the pilots, make it so. But is your life worth less over the heartland of America rather than over the Atlantic? It certainly is if you are on the low-cost regional carrier. If you are on such a plane headed to Denver and the engine is on fire, I am sure it is comforting to know that you saved 15 percent by scouring the Internet for the cheapest fare. Isn't it great to know that you have the newest, least-experienced, exhausted, starving young cockpit crew that this regional airline could find?

Did I say starving? Yeah, I did. Did you know that these regional crews can work for twelve to thirteen hours every day, flying five to eight legs a day, but their airline does not feel it's important enough to provide food for them? They are already on welfare wages, and now they have to find time and money while on the ground for twenty-five minutes to simply nourish themselves. It's a sad state of affairs. Remember, you bought the cheapest ticket.

HURRAY FOR HIGH GAS PRICES!
(SDL)

This post was published in June 2007, when the average price of regular gasoline in the U.S. was $2.80 per gallon, having risen dramatically in previous months. A year later, the price would hit $4. As of this writing (January 2015), the price has fallen all the way to $2.06 per gallon. So even without adjusting for inflation, gas *is 26 percent cheaper now than when this post was written. The federal gas tax, meanwhile, hasn't been raised since 1993.*

For a long time I have felt the price of gasoline in the United States was way too low. Pretty much all economists believe this, and also believe therefore that the gas tax should be raised substantially.

The reason we need high gas taxes is that there are all sorts of costs associated with my driving that I don't pay—someone else pays them. This is what economists call a "negative externality." Because I don't pay the full costs of my driving, I drive too much. Ideally, the government could correct this problem through a gas tax that aligns my own private incentive to drive with the social costs of driving.

Three possible externalities associated with driving are the following:

a. My driving increases congestion for other drivers.
b. I might crash into other cars or pedestrians.
c. My driving contributes to global warming.

If you had to guess, which of those three considerations provides the strongest justification for a bigger tax on gasoline?

The answer, at least based on the evidence I could find, may surprise you.

The most obvious one is congestion. Traffic jams are a direct consequence of too many cars on the road. If you took some cars away, the remaining drivers could get places much faster. From Wikipedia's page on traffic congestion:

The Texas Transportation Institute estimates that in 2000 the 75 largest metropolitan areas experienced 3.6 billion vehicle-hours of delay, resulting in 5.7 billion US gallons (21.6 billion liters) in wasted fuel and $67.5 billion in lost productivity, or about 0.7% of the nation's GDP.

This particular study doesn't tell us what we really need to know for estimating how big the gas tax should be. (We want to know how much adding one driver to the mix affects lost productivity.) But it does get to the point that, as a commuter, I'm better off if you decide to call in sick today.

A more subtle benefit of having fewer drivers is that there would be fewer crashes. Aaron Edlin and Pinar Mandic, in a paper I was proud to publish in the *Journal of Political Economy*, argue convincingly that each extra driver raises the insurance costs of other drivers by about $2,000. Their key point is that if my car is not there to crash into, maybe a crash never happens. They conclude that the appropriate tax would generate $220 billion annually. So, if they are right, reducing the number of crashes is a more important justification for a gas tax than reducing congestion. I'm not sure I believe this; it certainly is a result I never would have guessed to be true.

How about global warming? Every gallon of gas I burn releases carbon into the atmosphere, presumably speeding global warming. If you can believe Wikipedia's entry on the carbon tax, the social cost of a ton of carbon put into the atmosphere is about forty-three dollars. (Obviously there is a huge standard of error on this number, but let's just run with it.) If that number is right, then the gas tax needed to offset the global warming effect is about twelve cents per gallon. According to a National Academy of Sciences report, American motor vehicles burn about 160 billion gallons of gasoline and diesel each year. At twelve cents a gallon, that implies a $20 billion global warming externality. So relative to reducing congestion and lowering the number of accidents, fighting global warming is a distant third in terms of reasons to raise the gas tax. (Not that $20 billion is a small number; it just highlights how high the costs are from congestion and accidents.)

Combining all these numbers, along with the other reasons why we should raise the gas tax (e.g., wear and tear on roads), it seems easy to justify raising the gas tax by at least one dollar per gallon. In 2002 (the year I could easily find data for), the average tax was forty-two cents per gallon, or maybe only one-third of what it should be.

High gas prices act just like taxes, except that they are more transitory and the extra revenue goes to oil producers, refiners, and distributors instead of to the government.

My view is that, rather than bemoaning the high price of gas, we should be celebrating it. And, if any presidential candidate should come out in favor of a one-dollar-per-gallon tax on gas, vote for that candidate.

One hidden consequence of high gas prices: they lead to more traffic fatalities as drivers opt for smaller, fuel-efficient cars—and, increasingly, motorcycles. A 2014 study in the journal Injury Prevention *found that in California alone, a thirty-cent-per-gallon rise in gas prices led to an extra eight hundred motorbike-related deaths over a nine-year period.*

Chapter 4
Contested

*Every time we write a book, our publisher prints up a bunch of swag—T-shirts, posters, etc.—to use as promotion. They send us a few boxes, which inevitably wind up in a closet. One day we were thinking: How can we give this stuff away to people who might actually want it? That's when we decided to run our first blog contest, with the winner getting a piece of swag. These contests were so much fun—our blog readers are extraordinarily ingenious— that we held dozens of them. Here are a few of our favorites.*

WHAT IS THE MOST ADDICTIVE THING IN THE WORLD?
(SDL)

I was talking with my colleague and friend Gary Becker a while back about addiction. Among his many other accomplishments, for which he has won a Nobel, Becker introduced the idea of rational addiction.

When he told me his opinion as to the most addictive good, I was initially surprised and skeptical. On further reflection, I believe he is right.

So here is the quiz: What does Gary Becker think is the most addictive thing on earth?

THE FOLLOWING DAY . . .

More than six hundred readers took a shot at guessing what Gary Becker thinks is the most addictive thing on earth.

Lots of folks said things like crack and caffeine, but do you really think I'm going to offer a blog quiz with an obvious answer?

While not the answer I was looking for, there was something poetic about Deb's guess:

A yawn. A smile. Salt.

Before I give the answer, it is worth thinking about what it means for a good to be addictive. At least the way I think about it, an addictive thing has the following characteristics:

1. Once you start consuming it, you want to consume more and more of it.

2. Over time you build up a tolerance to it; i.e., you get less enjoyment out of consuming a fixed amount of it.

3. Pursuit of that good leads you to sacrifice everything else in your life to get it, potentially leading you to do ridiculous things to try to get the good.

4. There is a period of withdrawal when you stop consuming the good.

No doubt alcohol and crack cocaine fit that description well. In Becker's view, however, there is something even more addictive than substances: people.

When he first said this, it sounded kind of crazy to me. What does it mean to say that people are addictive?

Then I thought more about it, and I think he is right. Falling in love is the ultimate addiction. There is no question that in the early stages of attraction, spending a little bit of time with someone makes you desperately want more. Infatuation can be all-encompassing, and people will do anything to make a relationship blossom. They will risk everything and often end up looking utterly foolish. Once someone is in a relationship, however, the utility he or she derives from time with the beloved diminishes. The heady excitement of courtship gives way to something much more mundane. Even if a relationship isn't that good, for at least one of the parties there is a painful withdrawal period.

To get the exact answer I was looking for took until comment number 343, when Bobo responded "Other People." Many others were close. Jeff (comment 13) said "Society or human companionship." Laura (comment 47) said "Love."

I'll declare all three of them winners.

THE UNINTENDED CONSEQUENCES OF A TWITTER CONTEST
(SJD)

The other day, we woke up to realize that we were about hit our four-hundred-thousandth Twitter follower. So we put out the following tweet, offering some Freakonomics swag as a reward:

@freakonomics
We're at 399,987 Twitter followers. Thanks everyone!
Follower #400,000 will get Freakonomics swag!

Innocent enough, no?

But we had walked right into an incentive trap.

We monitored our Twitter status in order to identify the four-hundred-thousandth follower. It happened very fast, as new followers were signing up at what seemed to be a rate of five or six per second. So we counted carefully and, voilà, found our winner:

@freakonomics
@emeganboggs You're our 400,000th Twitter follower! Congrats! Contest is over, thanks everyone!

But then, when we went back to our main Twitter page, we found that we were *below* the four-hundred-thousand mark, by quite a bit. In fact, we had fewer followers after the contest than before.

So what happened?

If you're a Twitter pro, you've probably already figured it out. Our offer of swag created an incentive to unfollow and then refollow our feed. Appropriately enough, our followers informed us, and promptly:

@GuinevereXandra
@freakonomics isn't my incentive then to unfollow and refollow and repeat until i get to 400,000?

@Schrodert
@freakonomics And the de-follows into re-follows begin!

@Keyes
@freakonomics Hah, there go twenty of your followers. Like the twitter version of those penny-a-bid auctions.

@ChaseRoper
@freakonomics you just created an incentive for followers to unfollow and try to re-follow in order to be #400K.

I wish I could say this was a clever experiment but in fact it was simply a good lesson in Twitter incentives. So the person we thought was our four-hundred-thousandth follower, @emeganboggs, wasn't. We'll still send her some swag, but we'll also send some to a couple other people who actually were around the four hundred thousand mark. Even if they did unfollow us to get there :-). Thanks to everyone for a fun day on Twitter and yet another good lesson in unintended consequences.

CONTEST: A SIX-WORD MOTTO FOR THE U.S.?
(SJD)

Inspired by a recent trip to London and a (New York) *Times* article about England's reluctant search for a national motto (suggestions include "No Motto Please, We're British" and "One Mighty Empire, Slightly Used"), as well as by a new book on six-word memoirs for which I wrote a piece ("On the seventh word, he rested"), I invite you all to attempt the following:

Write a six-word motto for the U.S. of A.

Foreign players are most welcome. Feel free to punctuate your motto liberally (or, if you will, conservatively); for instance: "Battered? A bit. Beaten? Puhleeze. Onward!"

TWO WEEKS LATER . . .

Your response to our motto contest was quite strong, with more than 1,200 replies to date. Anyone looking for a good snapshot of public sentiment during this most interesting election year [2008] would do well to scroll through the comments: they are pretty damn illuminating, and not remotely sanguine.

The earliest comments tended to lean fairly hard to the left. Then, apparently because the contest was picked up on some right-leaning blogs, a long round of corrective mottoes came pouring in. Upon entering this fray, a cynic might give our motto contest the following motto:

Leftists Whine; Rightists Parry; Bedlam Accomplished

Or perhaps this:

Dead Split Between Patriots and Hatriots

Considering that this blog at least occasionally concerns itself with economics, I was surprised there weren't more suggestions having to do with free markets, maybe something like . . .

Creative Destruction at Its Very Finest

In the end, there were so many good, thoughtful, funny, heartfelt, and nasty suggestions that it was self-evidently beyond our ability to just pick a winner. So we narrowed the entries to the following five finalists. Please vote for your choice below, and the motto with the most votes within forty-eight hours will be declared the winner.

1. The Most Gentle Empire So Far

2. You Should See the Other Guy

3. Caution! Experiment in Progress Since 1776

4. Just Like Canada, with Better Bacon

5. Our Worst Critics Prefer to Stay

A WEEK LATER . . .

As promised, we tallied your votes for a new six-word motto for the U.S. The winner was clear:

Our Worst Critics Prefer to Stay (194 votes)

Here are the runners-up:

Caution! Experiment in Progress Since 1776 (134)

The Most Gentle Empire So Far (64) votes

You Should See the Other Guy (38)

Just Like Canada, with Better Bacon (18)

I applaud your choice of winner, and I especially applaud "edholston," who wrote the motto. "Our Worst Critics Prefer to Stay" is, while perhaps not outright uplifting, a wonderfully concise acknowledgment of the paradox that a capitalist democracy inevitably is: a place that is often well worth complaining about, and which allows you to complain as loudly as you wish.

It seems a small reward to get just a piece of Freakonomics swag for such a mighty task as writing a new motto for the United States, but that is all we have to offer. That, and our thanks—to Ed, and to all of you who participated.

Now, who among you can see about actually getting this motto adopted?

Chapter 5

How to Be Scared of the Wrong Thing

In SuperFreakonomics, we identified one of the most dangerous activities any person can engage in: walking drunk. Seriously. The data show that walking one mile drunk is eight times more dangerous than driving one mile drunk. But mostly people just laughed and ignored us. When it comes to evaluating risks, people stink for all sorts of reasons—from cognitive biases to the media's emphasis on rare events. Over the years that has generated blog fodder on subjects as diverse as the fear of strangers, running out of oil, and horseback riding.

WHOA NELLIE

(SJD)

Matthew Broderick recently broke his collarbone while riding a horse. This makes Broderick the fourth or fifth person I've heard about in recent months who was injured while riding a horse. This got me to thinking: How dangerous is horseback riding, especially compared to, say, riding a motorcycle?

A quick Google search turns up a 1990 CDC report: "Each year in the United States, an estimated 30 million persons ride horses. The rate of serious injury per number of riding hours is estimated to be higher for horseback riders than for motorcyclists and automobile racers."

Interestingly, the people who get hurt riding horses are often under the influence of alcohol, just like the people who get hurt (and hurt others) while driving motor vehicles.

So why don't we hear about all this horseback danger? I have a few guesses:

1. A lot of horse accidents occur on private property, and involve just one person.

2. Such accidents probably tend to not generate police reports, as a motorcycle or drag-racing accident inevitably would.

3. The kind of people who might typically call attention to unsafe activities like horses more than they like motorcycles.

4. A big motorcycle accident is more likely to make the evening news than a horse-riding

accident—unless, or course, the victim of the horse-riding accident is a Matthew Broderick or a Christopher Reeve.

I may be wrong on this, but I don't recall that Reeve's tragic accident was taken as a call to ban or regulate horseback riding—whereas when Ben Roethlisberger, e.g., was injured while riding his motorcycle without a helmet, all the discussion was about the foolishness of his act. I'm not saying Big Ben wasn't being foolish; but, as a Steelers fan, I guess I'm glad he wasn't riding a horse.

WHAT THE SECRETARY OF TRANSPORTATION HAS TO SAY ABOUT MY CAR-SEAT RESEARCH
(SDL)

On his official government blog, U.S. secretary of transportation Ray LaHood dismissed my research on child safety seats. This research found that car seats are no better than seat belts at reducing fatalities or serious injuries among children aged two to six; it was based on nearly thirty years' worth of data from the U.S. Fatality Analysis Reporting System as well as from crash tests that Dubner and I commissioned.

My favorite quote from the secretary:

"Now, if you want to slice up the data to be provocative, have at it. As a grandfather and as secretary of an agency whose number one mission is safety, I don't have that luxury."

Reading the secretary's blog post, I am struck by just how differently he is reacting to a challenge than Arne Duncan did when I first told him about my work on teacher cheating. Duncan, now the U.S. secretary of education, was in charge of the Chicago public schools at the time. I expected Duncan to do what LaHood did: dismiss the findings, circle the wagons, etc. But Duncan surprised me. He said that all he cared about was making sure the children were learning as much as possible, and teacher cheating was getting in the way of that. He invited me into a dialogue, and we ultimately made a difference.

If the ultimate goal in this case is really child safety, here's what LaHood might have written on his blog:

For a long time, we've been relying on car seats to keep our children safe. The existing academic literature up until recently confirmed the view that car seats are very successful in that goal. But in a series of papers in peer-reviewed journals, Steven Levitt and his co-authors have challenged that view using three different data sets collected by the Department of Transportation, as well as other data sets. I'm no data expert, and I have an agency to run, so I don't have the luxury of analyzing the data myself. But I am a grandfather and my agency's number one mission is safety, so I've asked the researchers in my agency to do the following:

1. Take a close look at the data sets we collect here in my agency, which are the basis for Levitt's work. Is it really the case that in these data there is little or no evidence that car seats outperform adult seat belts in protecting children ages two and up? Our benchmark for measuring the effectiveness of car seats has always been versus children who are unrestrained. Maybe we need to rethink this going forward?

2. Demand that the physicians at the Children's Hospital of Philadelphia, who have repeatedly found that car seats work, make their data publicly available. It is my understanding that these physicians have refused to share their data with

Levitt, but in the interest of getting to the truth, other researchers should have the chance to review what they have done.

3. Carry out a series of tests using crash-test dummies to determine whether adult seat belts do indeed pass all government crash-test requirements. In *SuperFreakonomics,* Levitt and Dubner report on their findings with a very small sample of tests; we need much more evidence on the data.

4. Try to understand why, even *after thirty years,* the great majority of car seats are still not properly installed. After all this time, can we really blame it on the parents, or should the blame be put elsewhere?

5. After exploring all these issues, let's figure out the truth, and let's use it to guide public policy.

And if Secretary LaHood has any interest in pursuing any of these avenues, I stand at the ready to offer whatever help that I can.

Update: Secretary LaHood never did take me up on my offer to help.

SECURITY OVERKILL,
DIAPER-CHANGING EDITION
(SJD)

I've been thinking a bit lately about security overkill. This includes not just the notion of "security theater," but the many instances in which someone places a layer of security between me and my everyday activities with no apparent benefit.

My bank, for instance, would surely argue that its many and various anti-fraud measures are valuable. But in truth, they are a) meant to protect the bank, not me; and b) cumbersome to the point of ridiculous. It's gotten to where I can predict which credit-card charge will trigger the bank's idiot algorithm and freeze my account because it didn't like the zip code where I used the card.

And security overkill has trickled down into the civilian world. When the class parents at my kids' school send out a parent contact list at the start of each school year, it comes via a password-protected Excel spreadsheet. Keep in mind that this list doesn't contain Social Security numbers or bank information—just names, addresses, and phone numbers of the kids' parents. I can imagine the day several months hence when someone actually needs to use the list and will find herself locked out by the long-forgotten password.

The most ridiculous example of security overkill I've run across recently was at Thirtieth Street Station, Philadelphia's main train terminal. I took a photo of something I saw in the men's room:

Yes, it's a diaper-changing station that has been fitted with a padlock. The handwritten message at the top says "see attendant for combination." I'm sure we could dream up some bad things that might happen on an unlocked diaper-changing tray, and I'm guessing as with most security overkill this was inspired by one anomalous event that scared the jeepers out of someone (or got that someone's lawyers involved). But still . . .

THE LATEST TERRORIST THREAT
(SDL)

The best strategy I have found for reducing the aggravation of security screening is to pretend I am a terrorist and think about where the weaknesses are in security, and how I might slip through. I think I figured out a way to get a gun or explosives into the White House during the George W. Bush administration. But I was only invited to the White House once, so I never got a chance to test my theory for real on a return visit.

Traveling to Ireland recently, I learned of a new anti-terror method. The security personnel in Dublin demand that you remove from your carry-on bag not only your laptop but another item that I hadn't previously known to be dangerous: your umbrella. For the life of me, I cannot think of what evil I would do with an umbrella—or more to the point, what evil I could do with an umbrella that would be prevented by having me take it out of my carry-on and put it directly on the conveyor belt. I asked the screener why umbrellas go directly on the belt, but her accent was quite thick so I couldn't understand her answer. I think I heard the word *poking* in there somewhere.

Learning about the possible dangers posed by umbrellas has dramatically reduced my utility. Now, every time I fly in the U.S., where the security treatment of umbrellas is so cavalier, I will spend the entire flight in fear that a rogue umbrella has made its way onto the plane.

One thing is for sure: if I ever see a passenger pull an umbrella out of her carry-on bag while a flight is airborne, I will tackle her first and ask questions later!

"PEAK OIL": WELCOME TO THE MEDIA'S NEW VERSION OF SHARK ATTACKS
(SDL)

This post was published on August 21, 2005. It would have been difficult to find anyone then willing to predict that ten years hence, technological advances in petroleum extraction would allow the U.S. to overtake Saudi Arabia as the biggest oil producer in the world. But that is exactly what happened.

A recent *New York Times Magazine* cover story, by Peter Maass, is about "peak oil." The idea behind peak oil is that the world has been on a path of increasing oil production for many years, and now we are about to peak and go into a situation where there are dwindling reserves, leading to triple-digit prices for a barrel of oil, an unparalleled worldwide depression, and as one oil-crash website puts it, "Civilization as we know it is coming to an end soon."

One might think that doomsday proponents would be chastened by the long history of people of their ilk being wrong: Nostradamus, Malthus, Paul Ehrlich, etc. Clearly they are not.

What most doomsday scenarios get wrong is the fundamental idea of economics: people respond to incentives. If the price of a good goes up, people demand less of it, the companies that make it figure out how to make more of it, and everyone tries to figure out how to produce substitutes for it. Add to that the march of technological innovation (like the green revolution, birth control, etc.). The end result: markets generally figure out how to deal with problems of supply and demand.

Which is exactly the situation with oil right now. I don't know much about world oil reserves. I'm not even necessarily arguing with their facts about how much the output from existing oil fields is going to decline, or that world demand for oil is increasing. But these changes in supply and demand are slow and gradual—a few percent each year. Markets have a way of dealing with situations like this: prices rise a little bit. That is not a catastrophe; it is a message that some things that used to be worth doing at low oil prices are no longer worth doing. Some people will switch from SUVs to hybrids, for instance. Maybe we'll be willing to build some nuclear power plants, or it will become worth it to put solar panels on more houses.

The *New York Times* article totally flubs the economics time and again. Here is one example from the article:

The consequences of an actual shortfall of supply would be immense. If consumption begins to exceed production by even a small amount, the price of a barrel of oil could soar to triple-digit levels. This, in turn, could bring on a global recession, a result of exorbitant prices for transport fuels and for products that rely on petrochemicals—which is to say, almost every product on the market. The impact on the American way of life would be profound: cars cannot be propelled by roof-borne windmills. The suburban and exurban lifestyles, hinged to two-car families and constant trips to work, school and Wal-Mart, might become unaffordable or, if gas rationing is imposed, impossible. Carpools would be the least imposing of many inconveniences; the cost of home heating would soar—assuming, of course, that climate-controlled habitats do not become just a fond memory.

If oil prices rise, consumers of oil will be (a little) worse off. But we are talking about needing to cut demand by a few percent a year. That doesn't mean putting windmills on cars; it means cutting out a few low-value trips. It doesn't mean abandoning North Dakota, it means keeping the thermostat a degree or two cooler in the winter.

A little later, the author writes:

The onset of triple-digit prices might seem a blessing for the Saudis—they would receive greater amounts of money for their increasingly scarce oil. But one popular misunderstanding about the Saudis—and about OPEC in general—is that high prices, no matter how high, are to their benefit.

Although oil costing more than $60 a barrel hasn't caused a global recession, that could still happen: it can take a while for high prices to have their ruinous impact. And the higher above $60 that prices rise, the more likely a recession will become. High oil prices are inflationary; they raise the cost of virtually everything—from gasoline to jet fuel to plastics and fertilizers—and that means people buy less and travel less, which means a drop-off in economic activity. So after a brief windfall for producers, oil prices would slide as recession sets in and once-voracious economies slow down, using less oil. Prices have collapsed before, and not so long ago: in 1998, oil fell to $10 a barrel after an untimely increase in OPEC production and a reduction in demand from Asia, which was suffering through a financial crash.

Oops, there goes the whole peak-oil argument. When the price rises, demand falls, and oil prices slide. What happened to the "end of the world as we know it"? Now we are back to ten-dollars-a-barrel oil. Without realizing it, the author just invoked basic economics to invalidate the entire premise of the article!

Just for good measure, he goes on to write:

High prices can have another unfortunate effect for producers. When crude costs $10 a barrel or even $30 a barrel, alternative fuels are prohibitively expensive. For example, Canada has vast amounts of tar sands that can be rendered into heavy oil, but the cost of doing so is quite high. Yet those tar sands and other alternatives, like bioethanol, hydrogen fuel cells and liquid fuel from natural gas or coal, become economically viable as the going rate for a barrel rises past, say, $40 or more, especially if consuming governments choose to offer their own incentives or subsidies. So even if high prices don't cause a recession, the Saudis risk losing market share to rivals into whose nonfundamentalist hands Americans would much prefer to channel their energy dollars.

As he notes, high prices lead people to develop substitutes. Which is exactly why we don't need to panic over peak oil in the first place.

So why do I compare peak oil to shark attacks? It is because shark attacks mostly stay about constant, but fear of them goes up sharply when the media decides to report on them. The same thing, I bet, will now happen with peak oil. I expect tons of copycat journalism stoking the fears of consumers about oil-induced catastrophe, even though nothing fundamental has changed in the oil outlook in the last decade.

BETTING ON PEAK OIL
(SDL)

John Tierney wrote a great *New York Times* column in response to Peter Maass's *Times* article about peak oil that I criticized. Tierney and the energy banker Matthew Simmons, who is the point man for the peak-oil team, made a $5,000 bet as to whether the price of oil in 2010 would be above or below $200 a barrel (adjusted for inflation to be in 2005 dollars).

The bet was designed in the spirit of the famous bet between Julian Simon and Paul Ehrlich, which the economist Simon won when the five commodities that Ehrlich said would rise in price actually fell substantially.

I am a betting man. And when I see that the NYMEX December 2011 crude oil future is priced under $60 a barrel, under $200 looks like a pretty good price to me! So I asked Simmons if he wanted any more action.

He was kind enough to write me back. As it turns out, I wasn't the first economist to offer him some more action. He declined to take my bet but he stuck to his guns in his belief that oil is priced way too cheaply, and that "real economic pricing will soon end almost a century of fantasy prices."

One thing that Simmons is definitely right about is that oil and gas are mighty cheap by volume compared to other things we consume. Imagine that a brilliant inventor came along and said he had invented a pill you could drop into a gallon of distilled water to turn it into gasoline. How much would you be willing to pay per pill? For most of the last fifty years, the answer is next to nothing, because a gallon of gas usually costs about the same as a gallon of distilled water.

But one place where I think Simmons's logic goes awry is that he seems to be arguing that because a gallon of gas is so valuable relative to say, a rickshaw driver, it should be as expensive as a rickshaw driver. In reasonably competitive markets, like the ones for gas and oil and presumably rickshaws, the determinant of price is how much it costs to supply the good, not how much consumers are willing to pay. That is because the supply of the good is close to perfectly elastic over some reasonable time horizon. If there were huge profits to be made at some price, firms will compete away the profit by lowering price. How much consumers like the good just determines the quantity consumed when supply is perfectly elastic. That is why water, oxygen, and sunshine—all incredibly valuable products—are virtually free to consumers: it is cheap or free to supply to them. And that is why we use a lot of gas and oil, but not many rickshaws at current prices.

If the cost of supplying oil suddenly jumped, then prices would certainly rise, more in the short run than the long run, as people figured out how to substitute away from using gas and oil. (Rickshaws, most likely, won't be the primary form of substitution, at least not in the U.S.) Whether we should care about "peak oil" boils down to: 1) Will the cost of supplying oil jump; 2) If it does jump, by how much; and 3) How elastic is the demand?

John Tierney won his bet: the year-average price in 2010 for a barrel of oil was eighty dollars, or seventy-one dollars adjusted to 2005 dollars. Sadly, Matthew Simmons died in August of that year, at age sixty-seven. "The colleagues handling his affairs reviewed the numbers," Tierney wrote, "and declared that Mr. Simmons's $5,000 should be awarded to me."

DOES OBESITY KILL?
(SJD)

There is so much noise these days about obesity that it can be hard to figure out what's important about the issue and what's not. To try to keep track, I sometimes divide the obesity issue into three questions.

1. Why has the U.S. obesity rate risen so much? Many, many answers to this question have been offered, most of them having to do with changes in diet and lifestyle (and, to some degree, the changing definition of *obese*). An interesting paper by the economists Shin-Yi Chou, Michael Grossman, and Henry Saffer sorts through many factors (including per capita number of restaurants, portion sizes and prices, etc.) and concludes—not surprisingly—that the spike in obesity mostly has to do with the widespread

availability of very cheap, very tasty food. They also find that a widespread decline in cigarette smoking has helped drive the obesity rate. This seems sensible, as nicotine is both a stimulant (which helps burn calories) and an appetite suppressant. But Jonathan Gruber and Michael Frakes have written a paper calling into doubt whether a decrease in smoking indeed causes weight gain.

2. How can obese people stop being obese? This, of course, is the question that sustains a multi-billion-dollar diet and exercise industry. A quick look at Amazon.com's top fifty books reveals just how badly people want to lose weight: there's *Intuitive Eating: A Revolutionary Program That Works; The Fat Smash Diet: The Last Diet You'll Ever Need;* and *Ultrametabolism: The Simple Plan for Automatic Weight Loss.* All these books make me think of the argument that every story in human history, from the Bible up through the most recent Superman movie, is built from one of seven dramatic templates. (FWIW, Superman and the Bible are plainly cut from the same template: baby Superman and baby Moses are both rescued from certain death,

sent off by their desperate parents in a rocket ship/wicker basket, and are then raised by an alien family but always remember the ways of their people and spend their lives fighting for justice.) This seven-template theory is even more true of diet books. They are pretty much all the same idea with some scrambled variables.

3. How dangerous is obesity? This is, to me, the toughest question of all. The conventional wisdom holds that obesity is like a huge wave that is just starting to break across the U.S., creating an endless swamp of medical and economic problems. But there is a growing sentiment that the panic over obesity may be as big a problem as obesity itself. Among the proponents of this view is Eric Oliver, a political scientist at the University of Chicago and the author of *Fat Politics: The Real Story Behind America's Obesity Epidemic.* Oliver argues that the obesity debate is rife with lies and misinformation. The book purports to show, as the jacket copy says, "how a handful of doctors, government bureaucrats, and health researchers, with financial backing from the drug and weight-loss industry, have campaigned to misclassify more than sixty million

Americans as 'overweight,' to inflate the health risks of being fat, and to promote the idea that obesity is a killer disease. In reviewing the scientific evidence, Oliver shows there is little proof either that obesity causes so many diseases and deaths or that losing weight makes people any healthier."

Well, even if Oliver is right, and putting aside for a moment Questions 1 and 2, obesity seems to be the culprit in at least twenty recent deaths. Last October, a tour boat carrying forty-seven elderly passengers sank on Lake George in upstate New York, and twenty of them died.

According to a National Transportation Safety Board report, this happened because the boat was badly overweight: the tour company used outdated passenger-weight standards to determine how many passengers the boat could safely carry. It wasn't over the passenger limit, but it was very much over the weight limit. And when the tourists crowded to one side of the boat to take in the view, disaster struck. According to *The New York Times*, the tour company had been using the old standard of 140 pounds per passenger, which the NTSB had already warned was no longer valid, and which Governor George Pataki has now updated for New York State, setting the new average-passenger weight at 174 pounds.

The legal wrangling has already been intense, with everybody looking to blame everybody else for the accident. The tour group has called the accident "an act of God." Others blame a company that modified the boat. Now it seems only logical that someone will step up to try to sue McDonald's for putting all those extra pounds on the passengers in the first place.

DANIEL KAHNEMAN ANSWERS YOUR QUESTIONS
(SDL)

One of the first times I met Danny Kahneman was over dinner, just after *SuperFreakonomics* was published. "I enjoyed your new book," Danny said. "It will change the future of the world." I beamed with pride. Danny, however, was not done speaking. "It will change the future of the world—and not for the better."

While I'm sure many people would agree, he was the only person who ever said it to my face!

If you don't know the name, Daniel Kahneman is the non-economist who has had the greatest influence on economics of any non-economist who ever lived. A psychologist, he's the only non-economist to win the

Nobel Prize in Economics for his pioneering work in behavioral economics. I don't think it would be an exaggeration to say that he is among the fifty most influential economic thinkers of all time, and among the ten most influential living economic thinkers.

In the years since that dinner, I've gotten to know Danny quite well. Every time I am with him, he teaches me something. His particular brilliance, I have decided, is being able to see what should be totally obvious, but somehow no one else manages to notice until he points it out.

Now he has written a fantastic book aimed at a popular audience: *Thinking, Fast and Slow.* It is a wonderfully engaging stroll through the world of behavioral economics—the kind of book that people are going to be talking about for a long, long time. Danny has generously agreed to answer questions from Freakonomics blog readers, which are paraphrased below. Here are his replies.

Q. A lot of research by you and others in the field proves that we often make irrational decisions—but what about research finding ways to be more rational? Have you tried this too?

A. Yes, of course, many have tried. I don't believe that self-help is likely to succeed, though it is a pretty good idea to slow down when the stakes are high. (And even the value of that advice has been questioned.) Improving decision making is more likely to work in organizations.

Q. Does your research say anything about risk-taking of the sort taken by the administrators at Penn State, who chose not to expose the sexual crimes of football coach Jerry Sandusky?

A. In such a case, the loss associated with bringing the scandal into the open now is large, immediate, and easy to imagine, whereas the disastrous consequences of procrastination are both vague and delayed. This is probably how many cover-up attempts begin. If people were certain that cover-ups would have very bad personal consequences (as happened in this case), we may see fewer cover-ups in future. From that point of view, the decisive reaction of the board of the university is likely to have beneficial consequences down the road.

Q. Levitt said you thought *SuperFreakonomics* would change the world for the worse. What did you mean by that?

A. It was a joking comment on the discussion of technological solutions to the global-warming problem in *SuperFreakonomics.* I thought that the favorable presentation of some solutions could suggest to readers that there is not much to worry about if the problem is easily solved. Not a serious disagreement.

Q. How can your research and writing help people make better decisions in health care, whether it's on the demand side or the supply side?

A. I don't believe that you can expect the choices of patients and providers to change without changing the situation in which they operate. The incentives of fee-for-service are powerful, and so is the social norm that health is priceless (especially when paid for by a third party). Where the psychology of behavior change and the nudges of behavioral economics come into play is in planning for a transition to a better system. The question that must be asked is "How can we make it easy for physicians and patients to change in the desired direction?," which is closely related to, "Why don't they already want the change?" Quite often, when you raise this question, you may discover that some inexpensive tweaks in the context will substantially change behavior. (For example, we know that people are more likely to pay their taxes if they believe that other people pay their taxes.)

Q. Can you address the relationship between happiness and satisfaction?

A. Yes, being happy (on average) in the moment and being satisfied retrospectively are not the same thing. People are most likely to be happy if they spend a lot of time with people they love, and most likely to be satisfied if they achieve conventional goals, such as high income and a stable marriage.

Q. Do you have any advice for guiding otherwise intelligent people to consider the legitimacy of scientific ideas or evidence that they happen to disagree with?

A. It is useful to distinguish the content of thoughts from the mechanisms of thinking. Some biases (e.g., preconceived notions, unscientific beliefs, specific stereotypes) are biases of content and are likely to be culture-bound. Other biases (e.g., the neglect of statistics, the neglect of ambiguity, the general fact that we are prone to stereotyping) are inevitable side effects of the operation of general-purpose psychological mechanisms.

Q. Is it possible that one barrier for women trying to work in male-dominated fields is that such an environment demands extra mental effort on behalf of the women?

A. Being self-conscious takes up mental capacity and is certainly not good for performance. Furthermore, the more self-conscious you are, the more likely you are to interpret (and sometimes misinterpret) the attitudes of others as gender-based, which is bound to make things worse. However, there is hope: self-consciousness is likely to diminish when you are in a stable environment, interacting with people you know well. The trend appears to be favorable: improving attitudes of men, rising representation of women in many male-dominated occupations, so the future is likely to be better than the past.

THE PERILS OF TECHNOLOGY, IPAD EDITION
(SJD)

These days, I read a lot of books on an iPad using the Kindle app. It is for the most part a very good experience, especially for recreational reading.

The other day, on vacation with the family, I came across a pitfall. I was reading the old football novel *North Dallas Forty*. It's pretty entertaining—especially the race stuff and drug stuff. As it happened, my nine-year-old daughter was curled up beside me reading her book, *The Doll People*, in a dead-tree version. She took at look at what I was reading. Her eyes immediately found a four-letter word.

"Hey," she said. "That's a bad word!"

"Yes," I said. "Yes it is."

And then, out of some childish parental instinct, I covered up the offending word with my thumb. What was I so scared of? I don't know what I was even trying to accomplish. She had already seen the word! Did my thumb have the power to make her unsee it? Even if it did, what would be the benefit?

As it happened, my thumb didn't just hide the word from her view; it touched the word on the screen—which, helpfully, pulls up a dictionary definition of the word:

VULGAR SLANG v. [trans.] 1 have sexual intercourse with (someone).

<SPECIAL USAGE> [intrans.] (of two people) have sexual intercourse.

2 ruin or damage (something).

Thank you, technology. You are indeed a double-edged sword. And it serves me right for being so f--- ing scared of my daughter seeing a curse word.

THIS IS WHAT I CALL BEING RISK-AVERSE
(SDL)

I found myself in a Las Vegas sports book the other day with good friend and economist John List. Since we both live in Chicago and have kids who play baseball, we thought it would be fun to bet some money on the Chicago White Sox. It would give us a reason to root for the White Sox, and give our kids a reason to open up the morning paper to see if the team had won.

We have no special information about the White Sox, no inside information. It was purely for consumption value.

If the sports book would give us a fair bet, i.e., the equivalent of a fifty-fifty coin toss, we would be willing to bet a lot because we aren't very risk-averse. I'd say we would have been willing to bet at least $10,000, probably even more.

But of course the sports book doesn't offer fair bets. On the particular bet we were looking at—how many games the White Sox would win over the course of the regular season—the sports book charges about an 8 percent vigorish, or commission. At that price, we decided, we were willing to bet $2,500. Eight percent of $2,500 is $200, so essentially we were willing to pay the sports book $200 in expectation to let us place this bet.

So we strolled up to the betting window and said we wanted $2,500 on the White Sox to win more than 84.5 games this year.

The lady behind the counter said the biggest bet we could make was $300.

What?!

We asked her why, and she called over a manager who told us the reason: the casino "didn't want to take too much risk on this kind of bet."

This casino is part of Caesars Entertainment, the largest casino company in the world, with annual revenues approaching $10 billion. And they aren't willing to let us pay them $200 to flip a coin for $2,500?

The next thing you know, the casino will tell me I can't lay $2,500 on "black" at the roulette table. After all, it is essentially the same gamble as our White Sox bet—a coin toss in which the casino gets better than fair odds.

This seems like a crazy way to run a business. It is especially surprising because Caesars is one of the few big businesses run by an economist, Gary Loveman, who has brought good economic thinking to many other aspects of the company's operations.

If I weren't an economist, running a sports book would be a pretty good job. I wonder if Caesars is accepting résumés?

FOUR REASONS WHY THE U.S. CRACKDOWN ON INTERNET POKER IS A MISTAKE
(SDL)

The U.S. government recently shut down the three major Internet poker sites for American players. Here are four reasons why this move makes no sense:

1. *Prohibitions that focus on punishing suppliers are largely ineffective. The prohibition of Internet poker is no exception.*

    When there is consumer demand for a good or service, it is extremely difficult to fight the problem through government punishments of suppliers. Illegal drugs are a good example. Americans want cocaine. Over the last forty years of the "War on Drugs," we have expended enormous amounts of resources locking up drug dealers. (Contrary to public opinion, the punishment of drug users has been relatively limited; by my estimates, 95 percent of the prison time served has been by drug sellers, as opposed to users.) Especially when the demand for a good is inelastic, squashing supply is ineffective. Making life difficult for incumbent suppliers entices new entrants who are eager to meet existing demand.

How do I know that the U.S. crackdown on Internet poker sites is ineffective? Within thirty minutes of my account being shut down on Full Tilt Poker, one of the big companies affected by the crackdown, I was able to start an account at a different, smaller poker site, depositing $500 via my credit card with no problem.

2. *Relative to the consumer surplus generated by online poker, the externalities caused are small. Government interventions should focus on cases where the opposite is true.*

Americans love poker. In a given year, Americans pay billions of dollars to be able to play the game online. I don't think I am exaggerating when I estimate that more than five million Americans have played poker online. Professional poker players are celebrities. The typical online poker player is not hurting anyone else, just like the typical moviegoer or sports enthusiast. There are, of course, gambling addicts; addicts impose costs on others. But the nature of Internet poker, with readily enforceable limits on how much money can be downloaded in a given period, is actually a much better environment for regulating addictive behavior than are poker casinos.

3. *From a moral perspective, it is inconsistent for the government to condone and profit from gambling on the one hand, while criminalizing private providers of Internet poker on the other.*

It would be understandable if, for reasons I disagree with, the government adopted a consistent stance against gambling of all sorts. But governments are enormous beneficiaries of gambling income, both through lotteries and sanctioned casinos. So there can be no moral high ground on the issue. I am certainly sympathetic to the government's desire to glean tax revenue from gambling activities. The right way to do that, however, is not a prohibition, but rather a regulatory framework in which governments take their cut of the action. For all parties involved, that sort of system is more efficient than the current approach.

4. *Even under the government's own laws, it would seem that there is little question that online poker should be legal.*

While I personally think the logic underlying the Unlawful Internet Gambling Enforcement Act, which governs online gambling, is deeply flawed, it is nonetheless the law of the land. Under the UIGEA, games of skill are exempted from the law, which is supposed to apply only to games of chance. So legally, whether online poker is legal comes down to the court's interpretation of whether poker is predominantly a game of skill. If you've ever played poker, it would seem self-evident that poker is a game of skill. If you need any further evidence, I recently co-wrote a paper with Tom Miles, a professor at the University of Chicago Law School, called "The Role of Skill Versus Luck in Poker." It uses data from the 2010 World Series of Poker to confirm what was already obvious.

THE COST OF FEARING STRANGERS
(SJD)

What do Bruce Pardo and Atif Irfan have in common?

In case you're not familiar with their names, let me rephrase:

What does the white guy who dressed up as Santa and killed his ex-wife and her family (and then committed suicide) have in common with the Muslim guy who got thrown off an AirTran flight on suspicion of terrorism?

The answer is that both of them had their intentions badly misread. The one who should have been scary to people who knew him wasn't; and the one who scared the people who didn't know him turned out to not be scary at all.

As we'll see below, this is a common pattern. But before going forward, let me first backtrack a bit.

Pardo was a churchgoer whom no one pegged as a homicidal maniac. "He's a totally different person from what you hear and see on the news for what he did," said a family friend. "I'm shocked, literally, I'm shocked. I can't believe that's actually the same guy."

Irfan, born in Detroit, is a tax attorney who lives with his family in Alexandria, Virginia. He was flying from Washington, D.C., to Florida with several members of his family for a religious retreat. He and his brother were reportedly discussing which are the "safest" seats on an airplane. "Other people heard them, misconstrued them," an AirTran spokesman told *The Washington Post*. "It just so happened these people were of Muslim faith and appearance. It escalated, it got out of hand, and everyone took precautions." The "precautions" involved removing all the Irfans from the plane and calling in the FBI to question them. They were promptly cleared by the FBI as definitely-not-terrorists, but AirTran still wouldn't fly them to Florida.

So which would you be more scared of: an American Muslim family you knew nothing about or the guy from your church who had just gone through a divorce?

As we've written before, most people are terrible at risk assessment. They tend to overstate the risk of dramatic and unlikely events at the expense of more common and boring (if equally devastating) events. A given person might fear a terrorist attack and mad cow disease more than anything in the world, whereas in fact she'd be better off fearing a heart attack (and therefore taking care of herself) or salmonella (and therefore washing her cutting board thoroughly).

Why do we fear the unknown more than the known? That's a larger question than I can answer here (not that I'm capable anyway), but it probably has to do with the heuristics—the shortcut guesses—our brains use to solve problems, and the fact that these heuristics rely on the information already stored in our memories.

And what gets stored away? Anomalies—the big, rare, "black swan" events that are so dramatic, so unpredictable, and perhaps world-changing, that they imprint themselves on our memories and con us into thinking of them as typical, or at least likely, whereas in fact they are extraordinarily rare.

Which brings us back to Bruce Pardo and Atif Irfan. The people who didn't seem to fear Pardo were friends and relatives. The people who did fear Irfan were strangers. Everyone got it backward. In general, we fear strangers way more than we should. Consider a few supporting pieces of evidence:

1. In the U.S., the proportion of murder victims who knew their assailants to victims killed by strangers is about 3 to 1.

2. Sixty-four percent of women who are raped know their attackers; and 61 percent of female victims of aggravated assault know their attackers. (Men, on the other hand, are more likely to be assaulted by a stranger.)

3. How about child abduction? Isn't that the classic stranger crime? A 2007 *Slate* article explains that of the missing children in one recent year, "203,900 were family abductions, 58,200 were nonfamily abductions, and only 115 were 'stereotypical kidnappings,' defined in one study as 'a nonfamily abduction perpetrated by a slight acquaintance or stranger in which a child is detained overnight, transported at least 50 miles, held for ransom, or abducted with the intent to keep the child permanently, or killed.'"

So the next time your brain insists on fearing strangers, try to tell it to cool out a bit. It's not that you necessarily need to insist that it fear your friends and family instead—unless, of course, you are friends with someone like Bernie Madoff. Let's not forget that the greatest financial fraud in history was committed primarily among friends. And with friends like that, who needs strangers?

Chapter 6

If You're Not Cheating, You're Not Trying

"Cheating may or may not be human nature," we wrote *in the first chapter of* Freakonomics, *"but it is certainly a prominent feature in just about every human endeavor. Cheating is a primordial economic act: getting more for less." That chapter was called "What Do Schoolteachers and Sumo Wrestlers Have in Common?" Over the ensuing ten years, we've had no trouble finding further evidence in support of this argument.*

CHEATING TO BE HOT
(SJD)

Are we too cynical?

I don't think so, but some people do. We routinely hear from readers who say it's a shame that we've called attention to so much deceit, trickery, and cheating among sumo wrestlers, schoolteachers, tax filers, and online daters. I could argue back and say, "Hey, we also called attention to people who don't cheat, like the office workers who put money in an 'honesty box' to pay for their bagels."

The point isn't that you can divide people into piles of bad people and good people, cheaters and non-cheaters. The point is that people's behavior is determined by how the incentives of a particular scenario are aligned.

So it was interesting to read Farhad Manjoo's article on *Salon* about a contest run by *FishbowlDC* to decide Washington's two hottest media folks. While agreeing that the winners were indeed a comely pair, Manjoo reports that the contest was a total rig job:

[The winners] Capps and Andrews acknowledge that they won only because their online friends—without their express encouragement, they both say—built software "bots" that voted thousands of times for each of them. The bots were distributed on Unfogged, a humorously wonky blog and discussion site popular with D.C. types, within a day of the poll's opening. If you downloaded and ran the software, your machine began tallying up votes for Capps and Andrews faster than a Diebold rigged for George W. Bush.

Which makes me say:

1. The stakes don't have to be very high for people to cheat.

2. When no punishment exists for cheating, it's pretty damn appealing.

3. We have been accused of stuffing a ballot box or two ourselves although there were no bots involved (that I know of).

4. Can anyone please point me in the direction of the Diebold folks who might have rigged those machines? It would be fun to talk to them!

WHY DO YOU LIE? THE PERILS OF SELF-REPORTING
(SJD)

I am always surprised at how easily, and cheaply, we humans lie.

Have you ever been in a conversation about, say, a particular book and been tempted to say you've read it even though you haven't?

I am guessing the answer is yes. But why would anyone bother to lie in such a low-stakes situation?

The book lie is what you might call a lie of reputation: you are concerned with what other people think of you. Of the many reasons that people lie, I have always thought that the lie of reputation is the most interesting—as opposed to a lie to gain advantage, to avoid trouble, to get out of an obligation, etc.

A new paper by César Martinelli and Susan W. Parker, called "Deception and Misreporting in a Social Program," offers some fascinating insights into lies of reputation. It takes advantage of a remarkably rich data set from the Mexican welfare program Oportunidades. It records the household goods that people say they have when they are applying for the program and it also records the household goods that are actually found in that household once the recipient's application has been accepted. Martinelli and Parker worked with data from more than one hundred thousand applicants, representing 10 percent of the applicants interviewed that year (2002).

It turned out that a lot of people underreported certain items that they thought might exclude them from getting benefits. Below is a list of underreported items followed by the percentage of recipients who owned that item but said they didn't:

Car (83 percent)
Truck (82 percent)
Video recorder (80 percent)
Satellite TV (74 percent)
Gas boiler (73 percent)
Phone (73 percent)
Washing machine (53 percent)

That's not very surprising: you might expect people to lie to gain the advantage of a welfare benefit. But here's the surprise. Below is a list of household items that were overreported—i.e., the items that applicants said they had but in fact did not (again, followed by percentages):

Toilet (39 percent)
Tap water (32 percent)
Gas stove (29 percent)
Concrete floor (25 percent)
Refrigerator (12 percent)

So four out of ten applicants without a toilet said they had one. Why?

Martinelli and Parker chalk it up to embarrassment, plain and simple. People who were desperately poor were also apparently desperate to not admit to a welfare clerk that they lived without a toilet or running water or even a concrete floor. This is one of the most amazing lies of reputation I can imagine.

It should be noted that there is a lot of incentive to lie to get into the Oportunidades program, for the cash benefit equals about 25 percent of the average applicant household's expenditures. Furthermore, the penalty for underreporting was not very strong: many of the people found to be underreporting goods like satellite TVs and trucks were not kicked out of the program. You could argue that the penalty for overreporting, meanwhile, was greater since it might mean being excluded from the program in the first place—which makes the overreporting even more costly.

The Martinelli-Parker paper may have broad implications for not only poverty programs but any kind of project where the data are self-reported. Think about a typical survey on drug use, sexual behavior, personal hygiene, voting preference, environmental behavior, etc. Here's what we once wrote, for instance, in an article about the lack of hand hygiene in hospitals:

> In one Australian medical study, doctors self-reported their hand-washing rate at 73 percent, whereas when these same doctors were observed, their actual rate was a paltry 9 percent.

We've also written about the topics that online daters are most likely to lie about and the risky business of election polling—especially when the issue of race is involved.

But as often as we or anyone else writes about the perils of self-reporting, the Martinelli-Parker paper really gives the whole topic a foundation to stand on. Not only does it deliver a surprising insight into why we lie, but it is also a sobering reminder to naturally distrust self-reported data—at least until some scientists enable us to peer into one another's minds and see what's really going on there.

HOW TO CHEAT THE MUMBAI TRAIN SYSTEM
(SJD)

A blogger named Ganesh Kulkarni discovered that the commuter trains of Mumbai serve six million passengers daily but the system isn't equipped to check everyone's ticket. Instead, Kulkarni writes, ticket agents conduct random ticket checks. This has given rise to a form of cheating that is elegantly called "ticketless travel." Although it's probably not very common to get busted for traveling ticketlessly, there is a significant fine if you are. And so, Kulkarni writes, one clever traveler has devised an insurance policy to make sure that ticketless travelers who are caught can lay off some of the expense.

Here's how it works. You pay five hundred rupees (about eleven dollars) to join an organization of fellow ticketless travelers. If you do get caught traveling without a ticket, you pay the fine and turn in your receipt to the ticketless-traveler organization, which refunds you 100 percent of the fine.

Don't you wish that everyone in society was as creative as the cheaters?

WHY DOES THE POST OFFICE DELIVER MAIL THAT HAS NO STAMP?
(SDL)

If you had asked me that question a week ago, I would have said with great certainty that the post office would not mail a letter without a stamp.

A few days ago, however, my daughter got a letter delivered in the mail. Where the stamp should have been, the sender had instead written "Exempt from postage: Guinness Book of World Records attempt."

The envelope contained a single sheet of paper, describing an attempt to set the record for the world's longest-running chain letter, along with instructions to pass this letter along to seven friends. The letter said that if we broke the chain, the Postal Service, which was monitoring the record attempt, would know that we were the individuals who ruined it for all of the people who had been part of the chain since 1991!

The simple arithmetic of chain letters guaranteed that somebody was lying.

A chain letter for which every recipient actually forwarded the letter to seven other people would quickly absorb every child in the world (seven raised to the power of ten is roughly the U.S. population). I did, however, give the sender credit for at least admitting this was a chain letter.

The thing that puzzled me was why the Postal Service was aiding and abetting this effort. It seemed bizarre, but at the same time lent credence to the endeavor. Maybe this really did have something to do with a world record bid.

A quick Google search, however, revealed that the Postal Service isn't condoning the chain mail. Actually, the explanation for why the letter got delivered without postage is even more interesting to me: apparently the automated mail-sorting machines fail to catch many letters that are missing a stamp.

On reflection, this does make sense—profit maximization requires setting the marginal cost of an action equal to the marginal benefit. If almost all letters have stamps, then the benefit of checking each one with 100 percent accuracy is infinitesimal, so it makes sense to let some unstamped letters through. (The same idea holds for catching people who don't pay their train fare.)

Now, I am curious to know exactly how lax the Postal Service is. I'm just about to drop something in the mail. Maybe I'll skip the stamp—though I suspect my tax return will make its way to the IRS, stamp or no stamp.

HERD MENTALITY? THE FREAKONOMICS OF BOARDING A BUS
(SJD)

A few days a week, I bring my daughter to nursery school on the East Side of Manhattan. We live on the West Side, and usually take the bus across town. It is a busy time of day. At the bus stop closest to our apartment (we'll call this Point A), there are often forty or fifty people waiting for the bus. This is largely because there is a subway stop right there; a lot of people take the train from uptown or downtown, then go aboveground to catch the crosstown bus.

I don't like crowds much in general (I know: What am I doing living in New York?), and I especially don't like fighting a crowd when I'm trying to cram onto a bus with my five-year-old daughter. Because there are so many people waiting for a bus at Point A, we have perhaps a 30 percent chance of getting aboard the first bus that stops there, and probably an 80 percent chance of getting aboard one of the first two buses that stop at Point A. It's that crowded.

As for getting a seat on the bus, we have perhaps a 10 percent chance of sitting down on either of the first two buses at Point A. It's not such a long ride across town, maybe fifteen minutes, but standing on a crowded bus in winter gear, my daughter's lunch getting smushed in her backpack, isn't the ideal way to start the day. Point A is so crowded that when eastbound passengers get off the bus at Point A, using the bus's back door, a bunch of people surge onto the bus via the back door, which means that a) they don't pay, since the paybox is up front, and b) they take room away from the people who are legitimately waiting at the head of the crowd to get on the bus.

So a while ago, we started walking a block west to catch the bus at what we'll call Point B. Point B is perhaps 250 yards west of Point A, and therefore 250 yards farther from our destination. But at Point B, where there is no subway stop, the lines are considerably shorter, and the buses arrive less crowded. At Point B, we have a 90 percent chance of getting aboard the first bus that arrives, and perhaps a 40 percent chance of getting a seat. To me, this seems well worth the effort and time of walking 250 yards.

Once we hit upon this solution, we haven't boarded a single bus at Point A. We get to sit; we get to listen to the iPod together (we both love Lily Allen, and I don't worry so much about the fresh parts since Lily's British accent renders them nearly indecipherable for Anya); we don't arrive with a smushed lunch.

But what I can't figure out is why so few (if any) bus passengers at Point A do what we do. To anyone standing at Point A morning after morning, the conditions are obviously bad. The conditions at Point B are clearly better since a) Point B is close enough to see with the naked eye, and b) the buses that arrive at Point A from Point B often have room on them, although only for the first ten or twenty passengers trying to board at Point A.

Personally, I am happy that more people at Point A don't go to Point B (which would make me have to consider boarding at Point C), but I don't understand why this is so. Here are a few possibilities:

1. Walking 250 yards doesn't seem like a worthwhile investment to improve a short, if miserable, experience.

2. Having just gotten off the subway, the Point A passengers are already broken in spirit and can't muster the energy to improve their commute.

3. Perhaps some Point A passengers simply never think about the existence of a Point B, or at least the conditions thereof.

4. There is a herd at Point A; people may say they don't like being part of a herd, but psychologically they are somehow comforted by it; they succumb to "herd mentality" and unthinkingly tag along—because if everyone else is doing it, it must be the thing to do.

Personally, I am persuaded that all four points may be valid in varying measures, and there are undoubtedly additional points to be made. But if I had to pick an outright winner, I'd say number four: the herd mentality. The more social science we learn, the more we realize that people, while treasuring their independence, are in fact drawn to herd behavior in almost every aspect of daily life. The good news is that once you realize this, you can exploit the herd mentality for your own benefit (as in boarding a bus), or for the public good, as in invoking peer pressure to increase vaccination rates.

AN EXPERIMENT FOR FAKE MEMOIRS
(SJD)

Why are there so many fake memoirs in the world? The latest is Margaret Seltzer's *Love and Consequences*. (I would link to its Amazon page but, alas, it no longer has an Amazon page.)

If you had written a memoir that was, say, 60 percent true, would you try to present it as a memoir or as a novel? If you were the editor of a memoir that you thought was 90 percent true, would you publish it as a memoir or as a novel?

Or maybe a better question is: What are the upsides of publishing such a book as a memoir instead of a novel? Here are a few possible answers:

1. A true story gets a lot more media coverage than a lifelike novel.

2. A true story generates more buzz in general, including potential film sales, lecture opportunities, etc.

3. The reader is engaged with the story on a more visceral level if a book is a memoir rather than fictional.

Every time a memoir is exposed as a fake, you hear people say, "Well, if it's such a good story, why didn't they just publish it as a novel instead?" But I think reasons one to three above, and maybe many more, incentivize authors, publishers, and others to favor the memoir over the novel.

With number three in mind, and having read recently about how an expensive sugar-pill placebo works better than a cheap sugar-pill placebo, I thought of a fun memoir/novel experiment. Here's what you'd do:

Take an unpublished manuscript that tells an intense and harrowing story from a first-person perspective. Something along the lines of *A Million Little Pieces* or *Love and Consequences*. Assemble a group of one hundred volunteers and randomly divide them in half. Give a copy of the manuscript to fifty of them with a cover letter describing the memoir they are about to read. Give a copy of the manuscript to the other fifty with a cover letter describing the novel they are about to read. In each case, write and attach an extensive questionnaire about the reader's reaction to the book. Sit back, let them read, and compile the results. Does the "memoir" truly beat the "novel"?

THE LATEST ENTRY INTO THE CHEATING HALL OF FAME
(SDL)

If you like cheating, you have to love British rugby player Tom Williams's ploy last week.

Apparently there is a rule in rugby, as in soccer, that once a substitution is made to take a player out of the match, that player can't return to the game. The exception to this rule is "blood injuries," in which case a player can come off until the bleeding is stopped and then return to play.

Tom Williams suffered just such a blood injury at a very critical moment of a recent match. I don't know anything about rugby, but his team was down by a point and they had some sort of drop-kicking specialist on the sidelines and it was the perfect time for him to come in and try a kick that would give the lead to Williams's team, the Harlequins.

The trouble began when Williams left the field looking a bit too happy, considering the large quantities of blood pouring from his mouth. One might have written this off to his being a rugby player, but apparently even rugby players get cross when smashed in the mouth. This led to an investigation. Eventually, television footage revealed that Williams had pulled a capsule of theatrical blood out of his sock and bitten into it in order to produce the faked injury.

A brilliant idea, but alas, in the end not only did Williams get suspended, but his substitute missed the kick and the Harlequins lost the game by a single point.

IS CHEATING GOOD FOR SPORTS?
(SJD)

That was the question I found myself asking while reading through the *Times* sports section in recent days. I understand that we are sort of between seasons here. The Super Bowl is over, baseball has yet to begin, the NBA is slogging through its long wintry slog, and the NHL—well, I just don't care much about hockey, sorry.

In any case, this is plainly not a peak time of year for professional sports. But still: it is amazing how many sports articles have nothing to do with the games themselves, but rather the cheating that surrounds the games. Andy Pettitte apologizes to his teammates and Yankees fans for using HGH, and reveals that his friendship with Roger Clemens is strained . . . Clemens pulls out of an ESPN event so he doesn't cause "a distraction" . . . there are drug-testing articles about Alex Rodriguez, Miguel Tejada, and Éric Gagné.

And that's just baseball! You can also read about Bill Belichick's denial of taping opponents' practices and the continuing tale of doping cyclists. There are a few NBA articles, too (though nothing lately about refs' gambling), and soccer (though nothing lately on match fixing), but by and large, the sports section that arrives each morning feels more like a cheating section.

Maybe, however, this is just how we like it. As much as we profess to like the games for the games' sake, perhaps cheating is part of the appeal, a natural extension of sport that people condemn on moral grounds but secretly embrace as what makes sports most compelling. For all the talk of how cheating "destroys the integrity of the game," maybe that's not true at all? Perhaps cheating actually adds a layer of interest—a cat-and-mouse element, a detective-story element—that complements the game? Or maybe cheating is just another facet of the win-at-all-costs drive that makes a great athlete great? As the famous sports adage goes: "If you're not cheating, you're not trying."

Also, we love to applaud cheaters who have confessed their ways. Pettitte, for instance, got a hero's welcome for talking about his HGH mistakes; Clemens, meanwhile, with every further denial seems to be soaking up ill will like a sponge. Just as the theological concept of the Resurrection is so powerful, and just as a harsh winter is followed by an insistent spring, I wonder if our interest in sport, too, springs eternal, not in spite of the cheating scandals but because of them?

SHOULD WE JUST LET THE TOUR DE FRANCE DOPERS DOPE AWAY?

(SJD)

Now that virtually every cyclist in the Tour de France has been booted for doping, is it time to consider a radical rethinking of the doping issue?

Is it time, perhaps, to come up with a pre-approved list of performance-enhancing agents and procedures, require the riders to accept full responsibility for whatever long-term physical and emotional damage these agents and procedures may produce, and let everyone ride on a relatively even keel without having to ban the leader every third day?

If the cyclists are already doping, why should we worry about their health? If the sport is already so gravely compromised, why should we pretend it hasn't been? After all, doping in the Tour is nothing new. According to an MSNBC .com article, it was cycling that introduced the sports world to doping:

> The history of modern doping began with the cycling craze of the 1890s and the six-day races that lasted from Monday morning to Saturday night. Extra caffeine, peppermint, cocaine and strychnine were added to the riders' black coffee. Brandy was

added to tea. Cyclists were given nitroglycerine to ease breathing after sprints. This was a dangerous business, since these substances were doled out without medical supervision.

HOW WE WOULD FIGHT STEROIDS IF WE REALLY MEANT IT
(SDL)

Aaron Zelinsky, a student at Yale Law School, has proposed an interesting three-prong anti-steroid strategy for Major League Baseball:

1. An independent laboratory stores urine and blood samples for all players, and tests these blood samples ten years, twenty years, and thirty years later using the most up-to-date technology available.

2. Player salaries are paid over a thirty-year interval.

3. A player's remaining salary would be voided entirely if a drug test ever came back positive.

I'm not sure about points two and three, but there is no question that point one is essential to any serious attempt to combat the use of illegal performance enhancers. The state of the art in performance enhancement is the best set of techniques that cannot be detected using current technology. So, by definition, the most sophisticated dopers will evade detection, unless they are unlucky or make a mistake.

The threat of future improvements in testing technology is the most potent weapon available in this fight, because the user can never know for certain that the doping he does today won't be simple to detect a decade from now. Retrospective testing of samples attributed to Lance Armstrong suggest that he used EPO, which was not detectable at the time. The circumstances surrounding this test were sort of murky (the identification of the sample as Armstrong's was indirect, and it was also unclear why these samples were being tested in the first place), so the Tour de France champion didn't pay the price (at the time) that he would have if formal testing at later intervals had been a standard policy.

The athletes most likely to be deterred by this sort of policy are the superstars who have the most to lose if their long-term legacy becomes tarnished. Presumably, it is doping by superstars that is of the greatest concern to fans.

Zelinsky has provided a measuring stick against which we can see how serious Major League Baseball, or any other sport, is about fighting illegal performance enhancers: if the league adopts a policy of storing blood and urine samples for future testing, it is serious. Otherwise, it is not serious.

HOW NOT TO CHEAT
(SDL)

Let's say you discover an old lamp and rub it, and out comes a genie offering to grant you a wish. You are greedy and devious, so you wish for the ability, whenever you play online poker, to see all the cards that the other players are holding. The genie grants your wish.

What would you do next?

If you were a total idiot, you would do exactly what some cheaters on the website Absolute Poker appear to have done recently. Playing at the very highest-stakes games, they allegedly played every hand as if they knew every card that the other players had. They folded hands at the end that no normal player would fold, and they raised with hands that were winners but would seem like losers if you didn't know the opponents' cards. They won money at a rate that was about a hundred times faster than a good player could reasonably expect to win.

Their play was so anomalous that, within a few days, they were discovered.

What did they do next?

Apparently, they played some more, now playing worse than anyone has ever played in the history of poker—in other words, trying to lose some of the money back so things didn't look so suspicious. One hand history shows that the players called a bet at the end when their two hole cards were two-three and had not paired the board: there was literally no hand that they could beat!

I don't know whether these cheating allegations are true, because all the information I am getting is third-hand. The poker players I've talked to all believe it to be true. Regardless, I bet these guys wish they had it to do over. If they had just been smart about it, they could have milked this gig forever, winning at reasonable rates. For the stakes they were playing, they could have gotten very rich, and their scheme would have been nearly undetectable.

(Note that I say nearly undetectable, because while that poker site probably never would have detected them, I am working with a different online poker site to develop a set of tools for catching cheaters. Even if these guys were careful, we would catch them.)

A FEW WEEKS LATER . . .

THE ABSOLUTE POKER CHEATING SCANDAL BLOWN WIDE OPEN
(SDL)

I recently blogged about allegations of cheating at an online poker site called Absolute Poker. While things looked awfully suspicious, there wasn't quite a smoking gun, and it was unclear exactly how the cheater might have cheated.

A combination of some incredible detective work by some poker players and an accidental (?) data leak by Absolute Poker have blown the scandal wide open.

The firsthand account can be found at 2+2 Poker Forum, and *The Washington Post* has followed up with an extensive report, but here's the short version:

Some opponents became suspicious of how a certain player was playing. He seemed to know what the opponents' hole cards were. The suspicious players provided examples of these hands, which were so outrageous that virtually all serious poker players were convinced that cheating had occurred. One of the players who'd been cheated requested that Absolute Poker provide hand histories from the tournament (which is standard practice for online sites). In this case, Absolute Poker "accidentally" did not send the usual hand histories, but instead sent a file that contained all sorts of private information that the poker site would never release.

The file contained every player's hole cards, observations of the tables, and even the IP addresses of every person playing. (I put "accidentally" in quotes because the mistake seems like too great a coincidence when you learn what followed.) I suspect that someone at Absolute knew about the cheating and how it happened, and was acting as a whistle-blower by sending these data. If that is the case, I hope whoever "accidentally" sent the file gets their proper hero's welcome in the end.

Then the poker players went to work analyzing the data—not the hand histories themselves, but other, more subtle information contained in the file. What these players-turned-detectives noticed was that, starting with the third hand of the tournament, there was an observer who watched every subsequent hand played by the cheater. (For those of you who don't know much about online poker, anyone who wants can observe a particular table, although, of course, the observers can't see any of the players' hole cards.) Interestingly, the cheater folded the first two hands before this observer showed up, then did not fold a single hand before the flop for the next twenty minutes, and then folded his hand pre-flop when another player had a pair of kings as hole cards! This sort of cheating went on throughout the tournament.

So the poker detectives turned their attention to this observer. They traced the observer's IP address and account name to the same set of servers that host Absolute Poker, and also, apparently, to a particular individual who seems to be employed by Absolute Poker! If all of this is correct, it shows exactly how the cheating would have transpired: an insider at the website had real-time access to all of the hole cards (it is not hard to believe that this capability would exist) and was relaying this information to an outside accomplice.

Online poker is a game of trust—players send their money to a site believing that they will be playing a fair game, and trusting that the site will send them their winnings. If there is even a little bit of uncertainty about either one of those factors, there is no good reason for a player to choose that site over the many close substitutes that exist. If I ran Absolute Poker, I would take a lesson from past corporate attempts at cover-ups, sacrifice the cheaters, and institute safeguards to prevent this ever happening again.

The real lesson of this all, however, is probably the following: guys who aren't that smart will figure out ways to cheat. And, with a little luck and the right data, folks who are a lot smarter will catch them doing it.

Update: According to The Washington Post, *Absolute Poker ultimately acknowledged it "found a breach in its software and is investigating." Shortly thereafter, the company "informs players that a high-ranking consultant in its Costa Rica office breached its software and spied on competitors' hands . . . But in a move that angers players, it refuses to identify the cheater or turn him over to authorities." Absolute Poker was later fined by a gaming commission but was allowed to keep its license. Meanwhile, according to the* Post, *"a new cheating scandal surfaces at sister site of Absolute Poker, UltimateBet.com." UltimateBet would later acknowledge insider cheating and pay more than $6 million in refunds—but, again, it was allowed to pay a fine and keep its license.*

TAX CHEATS OR TAX IDIOTS?
(SJD)

Neither Tom Daschle nor Nancy Killefer will be joining the Obama administration. Their nominations were both undone by their failure to pay taxes. Tim Geithner, meanwhile, was recently confirmed at treasury secretary despite his own tax failures.

Good God: What does it say about the U.S. tax code that people like Geithner, Daschle, and Killefer haven't properly paid their taxes?

(By "people like" them, I mean people who are smart and accomplished, have been through many application and vetting processes in their careers, and above all have reason to comply with tax paying.)

Here, let's make it a quiz:

a. If all three of them were intentionally cheating (and getting away with it until high-level scrutiny), then it's much too easy to cheat on taxes.

b. If all of them made honest mistakes, then the tax code simply isn't working.

c. If there's some combination of cheating and mistakes, then it's too easy to cheat and the tax code isn't working.

I'd vote for C. We once wrote a column about tax cheating which included this passage:

The first thing to remember is that the IRS doesn't write the tax code. The agency is quick to point its finger at the true villain: "In the United States, the Congress passes tax laws and requires taxpayers to comply," its mission statement says. "The IRS role is to help the large majority of compliant taxpayers with the tax law, while ensuring that the minority who are unwilling to comply pay their fair share."

So the IRS is like a street cop or, more precisely, the biggest fleet of street cops in the world, who are asked to enforce laws written by a few hundred people on behalf of a few hundred million people, a great many of whom find these laws too complex, too expensive and unfair.

Maybe the gross embarrassment over these high-profile tax failures will at least spur some tax-code sanity—like the Simple Return, promoted by the economist Austan Goolsbee, who has Obama's ear.

People like Daschle wouldn't fill out the Simple Return, but it might free up the IRS to catch tax violations before the Senate confirmation hearings flush them out.

HAVE D.C.'S "BEST SCHOOLS" BEEN CHEATING?
(SDL)

A USA Today investigation found what appears to be compelling evidence of teacher cheating in Washington, D.C., schools that were heralded as major successes due to test-score increases. The smoking gun: too many erasures resulting in answers being changed from wrong to right. The numbers are so extreme that they do seem to indicate that massive cheating occurred. Not surprisingly, the school district is none too eager to investigate—especially because teachers at the schools were given big bonuses as rewards for the test-score improvement. Though on Tuesday, acting D.C. schools chancellor Kaya Henderson did request a review.

When Brian Jacob and I investigated teacher cheating in Chicago schools, which we described in *Freakonomics*, we didn't use erasure analysis. Rather, we developed new tools for identifying strings of unlikely answers.

You might ask why we didn't use erasures, when it is such an obvious approach. The answer: unlike the D.C. schools, the Chicago schools did not farm out grading of the test exams to a third party. What got the D.C. schools in trouble is that the third party routinely analyzed erasure patterns. The internal group that scored the exams in Chicago did not routinely look for erasures; that was only done when there were suspicions about particular classrooms.

Conveniently, there was an acute shortage of storage space in the Chicago warehouse where the test forms were graded. This, of course, necessitated that all the actual test forms be destroyed and disposed of shortly after the test was administered.

Some teachers in D.C. are surely wishing there had been a similar lack of storage space in our nation's capital.

MAKING PROFITS FROM INCIVILITY ON THE ROADS
(SDL)

I hardly ever drive anymore since I moved close to where I work. So whenever I do, the incivility on the roads leaps out at me. People do things in cars they would never do in other settings. Honking. Swearing. Cutting to the front of the line. And that is just my sister. The other drivers are far meaner.

One obvious reason is that you don't have to live with the consequences for any length of time. If you cut in line at airport security, you will be in close proximity for quite some time to the people you insulted. With a car, you make a quick getaway. Making that getaway also means you are unlikely to be physically beaten, whereas giving someone the finger as you walk down the sidewalk has no such safety.

When I used to commute, there was one particular interchange where incivility ruled. (For those who know Chicago, it is where the Dan Ryan feeds into the Eisenhower.) There are two lanes when you exit the highway. One lane goes to another highway, the other goes to a surface street. Hardly anyone ever wants to go to that surface street. There can be a half-mile backup of cars waiting patiently to get on the highway, and about 20 percent of the drivers rudely and illegally cut in at the last second after pretending they are heading toward the surface street. Every honest person that waits in line is delayed fifteen minutes or more because of the cheaters.

Social scientists sometimes talk about the concept of "identity." It is the idea that you have a particular vision of the kind of person you are, and you feel awful when you do things that are out of line with that vision. That leads you to take actions that are seemingly not in your short-run best interest. In economics, George Akerlof and Rachel Kranton popularized the idea. I had read their papers, but in general have such a weak sense of identity that I never really understood what they were talking about. The first time I really got what they meant was when I realized that a key part of my identity was that I was not the kind of person who would cut in line to shorten my commute, even though it would be easy to do so, and seemed crazy to wait for fifteen minutes in this long line. But if I were to cut in line, I would have to fundamentally rethink the kind of person I was.

The fact that I don't mind when my taxi driver cuts in these lines (actually, I kind of enjoy it) probably shows that I have a long ways to go in my moral development.

All this is actually just a rambling prelude to my main point. I was in New York City the other day and my taxi driver bypassed a long line of cars exiting the freeway to cut in at the last second. As usual, I enjoyed being an innocent bystander/beneficiary to this little crime. But what happened next was even more gratifying to the economist in me. A police officer was standing in the middle of the road, waving over to the shoulder every car that cut in line, where a second officer was handing out tickets like an assembly line. By my rough estimate, these two officers were giving out thirty tickets an hour at $115 a pop. At over $1,500 per officer per hour (assuming the tickets get paid), this was a fantastic moneymaking proposition for the city. And it nails just the right people. Speeding doesn't really hurt other people very much, except indirectly. So to my mind it makes much more sense to go directly after the mean-spirited behavior like cutting in line. This is very much in the spirit of New York police commissioner Bill Bratton's "broken windows" policing philosophy. I'm not sure it cuts down the number of cheaters on the roads in any fundamental way since the probability of getting caught remains vanishingly small. Still, the beauty of it is that 1) every driver that follows the rules feels a rush of glee over the rude drivers getting nailed; and 2) it is a very efficient way of taxing bad behavior.

So my policy recommendation to police departments across the country is to find the spots on the roads that lend themselves to this sort of policing and let the fun begin.

Chapter 7

But Is It Good for the Planet?

 *Raise your hand if you are in favor of wasting natural resources, wiping out wildlife, and killing off the best planet ever. Just as we thought—not a lot of hands in the air. Therefore, just about any idea to protect the environment is considered a good idea. But the numbers often tell a different story.*

IS THE ENDANGERED SPECIES ACT BAD FOR ENDANGERED SPECIES?
(SDL)

My colleague and co-author John List is one of the most prolific and influential economists around.

He's got a new working paper with Michael Margolis and Daniel Osgood that makes the surprising claim that the Endangered Species Act—which is designed to help endangered species—may actually harm them.

Why? The key intuition is that after a species is designated as endangered, a decision has to be made about the geographic areas that will be considered critical habitats for that species. An initial set of boundaries is made, after which there are public hearings, and eventually a final decision on what land will be protected. In the meantime, while this debate is ongoing, there are strong incentives for private parties to try to develop land that they fear they may in the future be prevented from developing by the endangered species status. So in the short run, destruction of habitat is likely to actually increase.

Based on this theory, List et al. analyze the data for the cactus ferruginous pygmy owl near Tucson, Arizona. Indeed, they find that land development speeds up substantially in the areas that are going to be designated critical habitats.

This result, combined with the economist Sam Peltzman's observation that only 39 of the 1,300 species put on the endangered species list have ever been removed, do not paint a very optimistic picture of the efficacy of the Endangered Species Act.

BE GREEN: DRIVE
(SDL)

When it comes to saving the environment, things are often not as simple as they seem at first blush.

Take, for instance, the debate about paper bags versus plastic bags. For a number of years, anyone who opted for plastic bags at the grocery store risked the scorn of environmentalists. Now it seems that the consensus has swung the other direction, once a more careful cost accounting is done.

The same sort of uncertainty hangs over the choice of disposable diapers vs. cloth diapers.

At least some choices are beyond reproach environmentally. It is clearly better for the environment if someone walks to the corner store instead of drives, right?

Now even this seemingly obvious conclusion is being called into question by Chris Goodall, via John Tierney's *New York Times* blog. And Goodall is no right-wing nut; he is an environmentalist and author of the book *How to Live a Low-Carbon Life*.

Tierney writes:

If you walk 1.5 miles, Mr. Goodall calculates, and replace those calories by drinking about a cup of milk, the greenhouse emissions connected with that milk (like methane from the dairy farm and carbon dioxide from the delivery truck) are just about equal to the emissions from a typical car making the same trip. And if there were two of you making the trip, then the car would definitely be the more planet-friendly way to go.

DO WE REALLY NEED A FEW BILLION LOCAVORES?
(SJD)

We made some ice cream at home last weekend. Someone had given one of the kids an ice cream maker a while ago and we finally got around to using it. We decided to make orange sherbet. It took a pretty long time and it didn't taste very good but the worst part was how expensive it was. We spent about twelve dollars on heavy cream, half-and-half, orange juice, and food coloring—the only ingredient we already had was sugar—to make a quart of ice cream. For the same price, we could have bought at least a gallon (four times the amount) of much better orange sherbet. In the end, we wound up throwing away about three-quarters of what we made. Which means we spent twelve dollars, not counting labor or electricity or capital costs (somebody bought the machine, even if we didn't) for roughly three scoops of lousy ice cream.

As we have written before, it is a curious fact of modern life that one person's labor is another's leisure. Every day there are millions of people who cook and sew and farm for a living—and there are millions more who cook (probably in nicer kitchens) and sew (or knit or crochet) and farm (or garden) because they love to do so. Is this sensible? If people are satisfying their preferences, who cares if it costs them twenty dollars to produce a single cherry tomato (or twelve dollars for a few scoops of ice cream)?

This is the question that came to mind the other day when we received an e-mail from a reader named Amy Kormendy:

> I e-mailed Michael Pollan recently to ask him this question, and nice guy that he is, he promptly answered "Good question, I don't really know" and suggested I pose it to you good folks:
>
> Wouldn't it be more resource-intensive for us all to raise our own food than if we paid an expert to raise lots of food that s/he could sell to us? Couldn't it therefore be more sustainable to purchase food from large professional producers?
>
> Some of Professor Pollan's advice seems to be that we would be better off as a society if we did

more for ourselves (especially growing our own food). But I can't help but think that the economies of scale and division of labor inherent in modern industrial agriculture would still render the greatest efficiencies in resource investment. The extra benefit of growing your own food only works out if you count the unquantifiables such as the sense of accomplishment, learning, exercise, suntan, etc.

I very much understand the locavore instinct. To eat locally grown food or, even better, food that you've grown yourself, seems as if it should be 1) more delicious; 2) more nutritious; 3) cheaper; and 4) better for the environment. But is it?

1. "Deliciousness" is subjective. But one obvious point is that no one person can grow or produce all the things she would like to eat. As a kid who grew up on a small farm, I can tell you that after I had my fill of corn and asparagus and raspberries, all I really wanted was a Big Mac.

2. There's a lot to be said for the nutritional value of homegrown food. But again, since one person can grow only so much variety, there are bound to be big nutritional gaps in her diet that will need to be filled in.

3. Is it cheaper to grow your own food? It's not impossible but, as my little ice cream story above illustrates, there are huge inefficiencies at work here. Pretend that instead of just me making ice cream last weekend, it was all one hundred people who live in my building. Now we've collectively spent $1,200 to each have a few scoops of ice cream. Let's say you decide to plant a big vegetable garden this year to save money. Now factor in everything you need to buy to make it happen—the seeds, fertilizer, sprout cups, twine, tools, etc.—along with the transportation costs and the opportunity cost. Are you sure you really saved money by growing your own zucchini and corn? And what if a thousand of your neighbors did the same? Or here's another, non-food example: building your own home from scratch versus buying a prefab home. With a site-built home, you need to invest in all the tools, material, labor, and transportation costs to make it happen, and the myriad inefficiencies of having dozens of workmen's pickup trucks retrace the same route hundreds of times all for the sake of erecting one family's home—whereas factory-built homes create the opportunity for huge efficiencies by bundling labor, materials, transportation, etc.

4. But growing your own food has to be good for the environment, right? Well, keeping in mind the transportation inefficiencies mentioned above, consider the "food miles" argument and a recent article in *Environmental Science and Technology* by Christopher L. Weber and H. Scott Matthews of Carnegie Mellon:

> We find that although food is transported long distances in general (1640 km delivery and 6760 km life-cycle supply chain on average) the GHG [greenhouse gas] emissions associated with food are dominated by the production phase, contributing 83% of the average U.S. household's 8.1 t CO2e/yr footprint for food consumption. Transportation as a whole represents only 11% of life-cycle GHG emissions, and final delivery from producer to retail contributes only 4%. Different food groups exhibit a large range in GHG-intensity; on average, red meat is around 150% more GHG-intensive than chicken or fish. Thus, we suggest that dietary shift can be a more effective means of lowering an average household's food-related climate footprint than "buying local." Shifting less than one day per week's worth of calories from red meat and

dairy products to chicken, fish, eggs, or a vegetable-based diet achieves more GHG reduction than buying all locally sourced food.

This is a pretty strong argument against the perceived environmental and economic benefits of locavore behavior—mostly because Weber and Matthews identify the fact that is nearly always overlooked in such arguments: specialization is ruthlessly efficient. Which means less transportation, lower prices—and, in most cases, far more variety, which in my book means more deliciousness and more nutrition. The same store where I blew twelve dollars on ice cream ingredients will happily sell me ice cream in many flavors, dietetic options, and price points.

Whereas I am now stuck with about 99 percent of the food coloring I bought, which will probably sit in the cupboard until I die (hopefully not soon).

GOING GREEN TO INCREASE PROFITS
(SDL)

One of the hottest topics among businesspeople is how to increase profits by being environmentally friendly. There are many ways to achieve this. At hotels, for instance, by not automatically washing towels during a guest's stay, the hotel saves both money and the environment. Green innovations can be featured in advertising campaigns to attract customers. Another potential benefit of "going green" is that it makes environmentally minded employees happy, increasing their loyalty to the firm.

A Berlin brothel has hit on another way to use environmental arguments to its benefit: price discrimination. As Mary MacPherson Lane writes in an AP article:

> The bordellos in the capital of Germany, where prostitution is legal, have seen business suffer with the global financial crisis. Patrons have become more frugal and there are fewer potential customers coming to the city for business trips and conferences.
>
> But Maison d'Envie has seen its business begin to return since it began offering the euro 5 ($7.50) discount in July . . .
>
> To qualify, customers must show the receptionist either a bicycle padlock key or proof they used public transit to get to the neighborhood. That knocks the price for 45 minutes in a room, for example, to euro 65 from euro 70.

Although the brothel says the reason for the price discount is that it wants to be environmentally conscious, it sure looks to me like the brothel is dressing up some good old-fashioned price discrimination arguments in a green disguise.

Customers who come by bus or bicycle are likely to have lower incomes and be more price-sensitive than those who arrive by car. If that is the case, the brothel would like to charge such customers lower prices than the richer ones. The difficulty is that, without a justifiable rationale, the rich customers would be angry if the brothel tried to charge them more (and indeed, how in general, would the brothel know who is rich?). The environmental argument gives the brothel cover for doing what it always wanted to do anyway.

SAVING THE RAIN FOREST ONE GLASS OF ORANGE JUICE AT A TIME
(SDL)

I was drinking Tropicana orange juice this morning. The company has a clever marketing campaign. If you go to its website and type in the code on the Tropicana carton, they will set aside one hundred square feet of rain forest to preserve on your behalf.

What's clever about this?

I think corporations do not exploit the opportunities to bundle consumption of their products with contributions to charity as much as they probably should. I have no quantitative evidence on this; it is just a hunch. Typically, though, these sorts of corporate offers come in the form of "We will donate 3 percent of profits to X." The share of profits is usually small, which doesn't make the corporation seem generous.

The beauty of the rain forest offer is that one hundred square feet seems like a lot. Once you think about it, it isn't really much at all, but it sounds big. And if you are used to thinking about prices of land in cities, one hundred square feet could be expensive.

By my rough calculations, where I live it would cost about $130 to buy one hundred square feet of land you could build on. Land is cheap in the Amazon, however. Some online sites say that for $100, they will set aside an acre of land in the Amazon for you.

So probably, the true cost to Tropicana of an acre of Amazon land is half of that, or fifty dollars. Given the number of square feet in an acre, I calculate that the land my daughter saved in the Amazon this morning was worth about eleven cents. When I asked my daughter how much she thought the land was worth, she said twenty dollars. When I asked a friend, he guessed five dollars. Whenever a company can give away something worth eleven cents that people think is worth five or twenty dollars, they are doing something right.

The most remarkable thing is that even after we figured out our eleven-cent contribution, we still felt good about the fact that we had saved a little patch of land as big as the room we were eating breakfast in.

HOW ABOUT THEM (WRAPPED) APPLES?
(JAMES MCWILLIAMS)

James McWilliams, an historian at Texas State University, has written memorably about food production, food politics—and, as seen in the two Freakonomics.com posts below—the intersection of food and the environment.

Food packaging seems like a straightforward problem with a straightforward solution: there's too much of it; it piles up in landfills; we should reduce it. These opinions are standard among environmentalists, many of whom have undertaken impassioned campaigns to shroud consumer goods—including food—in less and less plastic, cardboard, and aluminum.

But the matter is a bit more complex than it might seem. Consider why we use packaging in the first place. In addition to protecting food from its microbial surroundings, packaging significantly prolongs shelf life, which in turn improves the chances of the food actually being eaten.

According to the Cucumber Growers' Association, just 1.5 grams of plastic wrap extends a cuke's shelf life from three to fourteen days, all the while protecting it from "dirty hands." Another study found that apples packed in a shrink-wrapped tray cut down on fruit damage (and discard) by 27 percent. Similar numbers have been found for potatoes and grapes. Again, while it seems too simple a point to reiterate, it's often forgotten: the longer food lasts, the better chances there are of someone consuming it.

True, if we all produced our own food, sourced our diet locally, or tolerated bruised and rotting produce, prolonging shelf life wouldn't matter much. But the reality is decidedly otherwise. The vast majority of food moves globally, sits in grocery stores for extended periods, and spends days, weeks, or even years in our pantries. Thus, if you accept the fact that packaging is an unavoidable reality of our globalized food system, you must also be prepared to draw a few basic distinctions. (If you don't accept that fact, well, there's probably no point in reading further.)

First, when it comes to food waste, not all materials are created equal. Concerned consumers look at wrapped produce and frown upon the packaging, because it's the packaging that's most likely destined for a landfill. But if you take the packaging away and focus on the naked food itself, you have to realize that the food will be rotting a lot sooner than if it weren't packaged and, as a result, will be heading to the same place as the packaging: the landfill. Decaying food emits methane, a greenhouse gas that's more than twenty times more potent than carbon dioxide. Packaging—unless it's biodegradable—does not. If the landfill is connected to a methane digester, which in all likelihood it isn't, you can turn the methane into energy. Otherwise, it makes more sense to send the wrapping (rather than the food) into the environmentally incorrect grave.

Second, when it comes to saving energy and reducing greenhouse gas emissions, our behavior in the kitchen far outweighs the environmental impact of whatever packaging happens to surround the product. Consumers toss out vastly more pounds of food than we do packaging—about six times as much. One study estimates that U.S. consumers throw out about half the food they buy. In Great Britain, the Waste and Resource Action Programme (WRAP, funnily enough) claims that the energy saved from not wasting food at home would be the equivalent of removing "1 out of every 5 cars off the road." The *Independent* reports that discarding food produces three times the carbon dioxide as discarding food packaging.

All of which is to say: if you're truly eager to take on the waste inherent in our food systems, you'd be better off reforming your own habits at home—say, by buying more strategically, minimizing waste, and eating less— before taking on the institutional packaging practices of disembodied food distributers.

Finally, we could also have an impact by choosing foods that are packaged in a way that reduces waste at home. This point does not apply so much to produce, but a lot of goods are packaged to ensure that we use the entire product. They contain user-friendly features such as capacious openings (milk), transparent appearance (bagged salad), re-sealers (nuts), the ability to be turned upside down (ketchup), and smooth surfaces rather than grooved ones, where food can hide (yogurt). Seems bizarre, but it's possible that we waste more energy by not scraping the bottom of the barrel than we do by throwing out the barrel when we're done. Given the high cost of wasting food, the question of design might be more important than the question of necessity.

Waste is an inevitable outcome of production. As consumers, we should certainly see food packaging as a form of waste and seek increasingly responsible packaging solutions. At the same time, though, we must do so without resorting to pat calls to "reduce packaging." Doing so, it seems, could do more harm than good.

AGNOSTIC CARNIVORES AND GLOBAL WARMING: WHY ENVIROS GO AFTER COAL AND NOT COWS
(JAMES MCWILLIAMS)

There's not a single person who's done more to fight climate change than Bill McKibben. Through thoughtful books, ubiquitous magazine contributions, and, most notably, the founding of 350.org (an international non-profit dedicated to fighting global warming), McKibben has committed his life to saving the planet. For all the passion fueling his efforts, though, there's something weirdly amiss in his approach to reducing greenhouse gas emissions: neither he nor 350.org will actively promote a vegan diet.

Given the nature of our current discourse on climate change, this omission might not seem a problem. Vegans are still considered as sort of "out there," a fringe group of animal rights activists with pasty skin and protein issues. However, as a recent report from the World Preservation Foundation confirms, ignoring veganism in the fight against climate change is sort of like ignoring fast food in the fight against obesity. Forget ending dirty coal or natural gas pipelines. As the WPF report shows, veganism offers the single most effective path to reducing global climate change.

The evidence is powerful. Eating a vegan diet, according to the study, is seven times more effective at reducing emissions than eating a local meat-based diet. A global vegan diet (of conventional crops) would reduce dietary emissions by 87 percent, compared to a token 8 percent for "sustainable meat and dairy." In light of the fact that the overall environmental impact of livestock is greater than that of burning coal, natural gas, and crude oil, this 87 percent cut (94 percent if the plants were grown organically) would come pretty close to putting 350.org out of business, which I'm sure would make McKibben a happy man.

There's much more to consider. Many consumers think they can substitute chicken for beef and make a meaningful difference in their dietary footprint. Not so. According to a 2010 study cited in the WPF report, such a substitution would achieve a "net reduction in environmental impact" of 5 to 13 percent. When it comes to lowering the costs of mitigating climate change, the study shows that a diet devoid of ruminants would reduce the costs of fighting climate change by 50 percent; a vegan diet would do so by over 80 percent. Overall, the point seems pretty strong: global veganism could do more than any other single action to reduce GHG emissions.

So why is it that 350.org tells me (in an e-mail) that, while it's "pretty clear" that eating less meat is a good idea, "we don't really take official stances on issues like veganism"? Well, why the heck not?! Why would an environmental organization committed to reducing greenhouse gas emissions not officially oppose the largest cause of greenhouse gas emissions—the production of meat and meat-based products? It's baffling. And while I don't have a definite answer, I do have a few thoughts on the matter.

Part of the problem is that environmentalists, including McKibben himself, are generally agnostic about meat. A recent article McKibben wrote for *Orion* reveals an otherwise principled environmentalist going a bit loopy in the face of the meat question. The tone is uncharacteristically cute, even folksy, and it's entirely out of sync with the gravity of the environmental issues at stake. Moreover, his claim that "I Do Not Have a Cow in this Fight" is a rather astounding assessment coming from a person who is so dedicated to reducing global warming that he supposedly keeps his thermostat in the fifties all winter and eschews destination vacations for fear of running up his personal carbon debt. I'd think the man has every cow in the world in this fight.

So to the real question: How do we explain this agnosticism? The fact that McKibben recently traveled to the White House to oppose the construction of a natural gas pipeline (and got arrested in the process), provides a hint of an answer. I imagine that getting slammed in the clinker after protesting a massive pipeline project is a lot better for 350 .org's profile than staying at home, munching kale, and advising others to explore veganism. In this respect, the comparative beneficial impact of global veganism versus eliminating natural gas from Canadian tar sands matters none. What matters is grabbing a headline or two.

Hence the "problem" with veganism and environmentalism. Ever since *Silent Spring,* Rachel Carson's exposé of dangerous insecticides, modern environmentalism has depended on high-profile media moments to shore up the activist base. Veganism, however, hardly lends itself to this role. Although quietly empowering in its own way, going vegan is an act poorly suited to sensational publicity. Pipelines and other brute technological intrusions, by contrast, are not only crudely visible, but they provide us with clear victims, perpetrators, and a dark narrative of decline. I think this distinction explains much of McKibben's—not to mention the environmental movement's—wobbly stance on meat.

Another reason for the prevailing agnosticism on meat has to do with the comparative aesthetics of pipelines and pastures. When meat-eating environmentalists are hit with the livestock conundrum, they almost always respond by arguing that we must replace feedlot farming with rotational grazing. Just turn farm animals out to pasture, they say. Not surprisingly, this is exactly what McKibben argues in the *Orion* piece, claiming that "shifting from feedlot farming to rotational grazing is one of the few changes we could make that's on the same scale as the problem of global warming."

This all sounds well and good. But if the statistics in the WPF report are to be trusted, the environmental impacts of this alternative would be minimal. So why the drumbeat of support for rotational grazing? I would suggest that the underlying appeal in the pasture solution is something not so much calculated as irrational: pastured animals mimic, however imperfectly, symbiotic patterns that existed before humans arrived to muck things up. In this sense, rotational grazing supports one of the more appealing (if damaging) myths at the core of contemporary environmentalism: the notion that nature is more natural in the absence of human beings. Put differently, rotational grazing speaks powerfully to the aesthetics of environmentalism while confirming a bias against the built environment; a pipeline, not so much.

A final reason that McKibben, 350.org, and mainstream environmentalism remain agnostic about meat centers on the idea of personal agency. For most people, meat is essentially something we cook and eat. Naturally, it's much more than that. But for most consumers, meat is first and foremost a personal decision about what we put into our body. By contrast, what comes to mind when you envision an old coal-fired power plant? Many will conjure up sooty images of a degraded environment. In this respect, the coal-fired power plant symbolizes not a personal choice, or a direct source of pleasure, but an oppressive intrusion into our lives, leaving us feeling violated and powerless. Environmentalists, I would thus venture, go after coal rather than cows not because coal is necessarily more harmful to the environment (it appears not to be) but because cows mean meat, and meat, however wrongly, means freedom to pursue happiness.

I don't mean to downplay the impact of these factors. The visibility of pipelines, the romantic appeal of pastures, and the deep-seated belief that we can eat whatever we damn well shove into our mouths are no mean hurdles to overcome. But given that the documented power of veganism to directly confront global warming, and given the fact that emissions have only intensified alongside all efforts to lower them, I'd suggest McKibben, 350.org, and the environmental movement as a whole trade up their carnivorous agnosticism for a fire-and brimstone dose of vegan fundamentalism.

HEY BABY, IS THAT A PRIUS YOU'RE DRIVING?
(SJD)

Remember when keeping up with the Joneses meant buying a diamond-encrusted cigarette case? Such ostentatious displays of wealth during the Gilded Age prompted Thorstein Veblen to coin the term *conspicuous consumption*.

Conspicuous consumption has hardly gone away— what do you think bling is?—but now it's got a right-minded cousin: conspicuous conservation. Whereas conspicuous consumption is meant to signal how much green you've got, conspicuous conservation signals how green you are. Like carrying that "I'm not a plastic bag" bag, or installing solar panels on the side of your house facing the street—even if that happens to be the shady side.

We recently made a podcast episode about conspicuous conservation; it features a research paper written by Alison and Steve Sexton, a pair of Ph.D. economics candidates who happen to be twins (and who have economist parents, too). The paper is called "Conspicuous

Conservation: The Prius Effect and Willingness to Pay for Environmental Bona Fides."

Why single out the Toyota Prius? Here's how Steve Sexton explains it:

"The Honda Civic hybrid looks like a regular Honda Civic. The Ford Escape hybrid looks like a Ford Escape. And so, our hypothesis is that if the Prius looked like a Toyota Camry or a Toyota Corolla, it wouldn't be as popular as it is. And so what we set out to do in this paper is to test that empirically."

The question they really wanted to answer was this: How much value do people who lean green place on being seen leaning green? The Sextons found that the Prius's "green halo" was quite valuable to its owners— and, the greener the neighborhood, the more valuable the Prius is.

Chapter 8
Hit on 21

 One thing the two of us have in common is we never quite grew up. Levitt still clings to adolescent fantasies of being a professional golfer. Dubner still worships the Pittsburgh Steelers with the intensity of an eleven-year-old. And, somehow, we keep ending up together in Las Vegas.

I HOPE PHIL GORDON WINS
THE WORLD SERIES OF POKER
(SJD)

The main event of the World Series of Poker is just getting under way at the Rio in Las Vegas. Why do I want Phil Gordon to win?

It's not just because he's such a nice guy, or because he's so smart, or because of his philanthropic endeavors, or even because he's so tall.

It has to do with the game of Rock, Paper, Scissors, aka Rochambeau.

Levitt and I were in Vegas recently to do research with a bunch of world-class poker players. Part of that research included a sixty-four-player Rochambeau charity tournament that Phil Gordon organized, and which Annie Duke won.

One night, Gordon and his Full Tilt Poker pals threw a big party at Pure, the sleek nightclub at Caesars Palace. It was big and noisy and fun, and I had a long and interesting conversation with Phil Gordon about a number of things. In the end, talk turned to Rochambeau. Words were exchanged and suddenly there was a challenge—me against Gordon, head-to-head in Rochambeau, best of nine throws for $100.

Levitt held the money. Then Gordon, who is about eight inches taller than anyone I know, leans over into my face and says, "I'm starting with Rock."

And he did. I threw Scissors, so he beat me. Score: 1–0.

But I had something up my sleeve. I started the match throwing a Seamstress—i.e., a three-throw gambit of Scissors, then Scissors, then another Scissors. Gordon, after his initial Rock, threw a Paper, then another Paper. I was up 2–1.

Finally, on the fourth throw, Gordon threw a Scissors. But I had thrown my fourth Scissors in a row, which meant we tied on that throw, leaving the score at 2–1. That's when Gordon leaned into my face again and said, "You do know that you can throw something besides Scissors, right?"

But my four consecutive Scissors throws—let's call it a SuperSeamstress—seemed to have shaken him. He recovered to tie it up at 2–2 and took the lead briefly at 3–2, but I tied him, then went up 4–3. He managed to tie me at 4–4 but, never in doubt, I threw one more Scissors and beat him, 5–4. He looked pretty stunned. Poor guy. It turned out that he really hates to throw Scissors.

So why do I want him to win the WSOP? Not because I feel sorry for beating him. Now more than ever, I believe that Rochambeau is a game of luck, and I happened to get lucky against a guy who is a really good poker player.

No, the reason I want Gordon to win is simply so I can tell my grandchildren someday that I beat the WSOP champ at something, even if it was something as meaningless as Rock, Paper, Scissors.

A FEW MONTHS LATER . . .

VEGAS RULES
(SJD)

So Levitt and I were in Las Vegas this weekend, doing some research. (Seriously: it's for a *Times* column on

Super Bowl gambling.) We had a little downtime and decided to play blackjack. It was New Year's Eve, at Caesars Palace, about 9 P.M. We sat down at an empty table where the dealer, a nice young woman from Michigan, was very patient in teaching us the various fine points that neither of us knew and which indicated that we were both inexperienced. Keep one hand in your lap; e.g., when you want a card, just flick your cards twice on the felt. When you're standing, tuck one card under your chip(s). And so on.

At one point, Levitt kind of gasped. He had had twenty-one but somehow had asked for another card. The last card was a two. It wasn't that he didn't know how to play, or count; he was just distracted—talking to me, he'd later claim—and the dealer had seen him do something, or fail to do something else, that indicated he wanted another card. So here he was with four cards: a face card, a four, a seven, and a two. The dealer looked sympathetic. I vouched for Levitt, told her he wasn't an idiot and surely wouldn't have hit on twenty-one intentionally. She seemed to believe us. She said she'd call over her supervisor to see what could be done.

She called the supervisor's name over her shoulder. I could see the supervisor, and I could see that he couldn't hear her. Remember, this is a casino on New Year's Eve; it was pretty noisy. She keeps calling, and

I keep seeing that he's not hearing her, but she won't turn around to call him. That would mean turning her back on her table full of chips and, even if Levitt was dumb enough to hit on twenty-one, he presumably was smart enough to grab a bunch of chips and run. (Or maybe, she was thinking, he's actually dumb like a fox and used this hitting-on-twenty-one trick all the time, to get the dealer to turn her back on the table.)

Finally I went and got the supervisor. When he came over, the dealer explained the situation. He seemed to accept Levitt's explanation.

Then he looked at me. "Did you want the card?" he asked, meaning the two that Levitt drew.

"Well, now that I see it, sure I want it," I said. I had seventeen; I certainly wouldn't have hit on seventeen, but a two would give me a lovely nineteen.

"Here," he said, and gave me the two. "Happy New Year."

Then the dealer took a card and busted.

I don't know much about gambling, but I do know that the next time I'm in Vegas and feel compelled to play some blackjack, I'm going to Caesars.

And just so you don't think that Levitt really is a complete gambling idiot: the next day, we sat down at the sports book and he grabbed a *Daily Racing Form* and studied it for about ten minutes and then went up and placed a bet. He found a horse, going off at 7–2, that had never run a race. But he saw something that he liked. He bet the horse to win and win only. And then we watched the race on one of the jumbo screens. It took a good sixty seconds for his horse to settle into the gate—we thought it would be scratched—but then it got in and the gates opened and his horse led wire to wire. It was a good bit more impressive than his blackjack.

A FEW MONTHS LATER . . .

WORLD SERIES OF POKER UPDATE: LEVITT TIES RECORD THAT CAN NEVER BE BROKEN
(SDL)

I recently went to Vegas to play in my first World Series of Poker event. The game was no-limit hold 'em. Each player started with five thousand chips.

So what record did I tie? The record for the least number of pots won by a player in a WSOP event: zero. I played for almost two hours and did not win a single hand. I didn't even manage to steal the blinds once. Despite promising Phil Gordon seconds before the tournament that I would not let ace-queen be my undoing, I lost two big pots with those hole cards.

(Both times an ace came up on the flop; neither time did the opponent have an ace; both times I still lost.) I probably played them wrong both times.

The beauty of the WSOP is that there is always another event the next day. Maybe I'll give it another try tomorrow—there's nowhere to go but up.

THE NEXT DAY . . .

ONE CARD AWAY FROM THE FINAL TABLE AT THE WORLD SERIES OF POKER (SDL)

What a difference a day makes.

I blogged yesterday about my first foray into World Series of Poker action. It started and ended very badly, with me failing to win a single hand.

Who knows why I signed up for another day of punishment at the hands of the poker pros the very next day. The structure of this tournament was different: a shootout. That means that the ten players at a table play until one has all the chips. Then that player moves on to the next round. After two rounds of this, the field of nine hundred is whittled down to nine players who make the final table.

My pessimism was only enhanced when I discovered that David "the Dragon" Pham had been seated next to me. He has won over $5 million in poker tournaments, has two WSOP bracelets, and was the defending champion in this very event! My table of ten had at least five full-time poker professionals.

Amazingly, after some good luck, I emerged the winner five hours later.

I needed to win one more table to make it to the final table, which would provide me bragging rights for life. I was lucky to have lunch with Phil Gordon, probably the best poker teacher in the world. He explained something to me over lunch that is fundamental to good poker, and probably somewhat obvious, but I had never understood it. (It's too valuable an insight to give away for free here; you'll have to buy one of Phil's books.)

The combination of that insight and a lot of good cards had me rolling at the second table. Unfortunately, I had to knock out my friend Brandon Adams, one of the very best poker players in the world and a great writer as well. Brandon is a classic example of opportunity cost . . . he makes so much money playing poker that he will likely never finish his economics Ph.D. at Harvard.

I found myself with a chip lead as the table was reduced to just me and one opponent, Thomas Fuller. I built my lead up to about 2–1 after forty-five minutes. Then I lost a bunch when I had ace-king suited and probably played it completely wrong. That made our chip stacks about even.

Not long after was the hand that undid me. Fuller made a standard raise pre-flop. I called with king-seven. The flop came king-queen-eight, all of different suits. I bet 7,200 chips and he called. The turn card was a seven. There were now two clubs on the board. I checked, hoping he would raise me and then I could re-raise him. That is exactly what happened. He bet eight thousand and I re-raised.

Much to my surprise, he then re-raised me. What could he have? I was hoping he had a queen. But maybe he had K10, KJ, K8, AK, or even two pair. Still, I forged ahead. I re-raised him again. Then he pushed all-in! I figured I was beaten, but I called his all-in raise. I was stunned when he turned over a six and a nine. All he had was a straight draw. He was bluffing. There were only eight cards in the deck that would make him a winner. I had an 82 percent chance of winning that hand. If I won that hand, I had over 90 percent of the chips, and was virtually certain to make the final table. A five came on the river, he hit his straight, and the Cinderella story had come to an end. I was out.

I have to say, though, that even as antisocial as I am, I really enjoyed the ride. It was one of the best gambling experiences I've ever had. The morning after, however, I feel like I have a terrible hangover, despite the fact that I didn't have a drop of alcohol. I know myself well enough to know exactly what that hangover is all about. Like any "good" gambler, I don't care very much whether I win or lose, as long as there is more gambling action on the horizon. But when the gambling is over, the crash comes.

Today is the crash for me. No more WSOP. No more gambling for a while. "Just" a family trip to the Hoover Dam and a long plane ride back to Chicago.

And maybe a little teeny pick-six ticket at Hollywood Park.

WHY ISN'T BACKGAMMON MORE POPULAR?
(SJD)

I've mentioned in the past that I love backgammon. A reader recently wrote to ask whether Levitt and I ever play and, more important, why a game as great as backgammon isn't more popular.

Sadly, Levitt and I have never played. But it's the second part of the question that got me thinking. Why not indeed? Off the top of my head, I'd say:

- Well, it's not *so* unpopular, and there are those who say a renaissance is perhaps under way. My friend James Altucher and I have a running game (101-point matches) that we usually play in diners or restaurants, and almost inevitably a small crowd (or at least the server) will hang out to watch and talk about the game . . .

• That said, yes, it's a fringe game. Why? I'd say it's because too many people play it without gambling, or at least without using the doubling cube. Without the cube, a game that can be intricate and strategic can too easily become a boring dice race. Once you use the cube, especially with dollars attached to points, the game changes completely because the most exciting and difficult decisions have to do more with cube play than with checker play.

• Why is the game itself too often uninteresting? Don't get me wrong: I love playing backgammon. But the fact is that the choice set of moves is in fact quite small. That is, for many rolls, there's clearly one optimal move, or perhaps two that are nearly equal. So once you know those moves, the game is limited, and you need some stakes to make it interesting. Unlike, say, chess, where the options and strategies are far more diverse.

This last point, if arguable, got me to wondering: In what percent of backgammon turns would there seem to be clearly one optimal move—versus, for comparison, chess?

Since James is a superb chess player and also an excellent backgammon player (and a smart guy in general), I asked him. His answer is well worth sharing:

It's an interesting question. Let's define optimal first.

Let's say a program has an evaluation function (EV). Given a position, the EV returns a number from one to ten based on how good the position is for the person whose move it is. If it's a ten, the person with the move wants to get to that position. The EV is a function of various heuristics added up (how many people are on the center, how many pips I'm ahead in the race, how many slots I control, how many loose pieces I have, etc.). When it's my turn, the computer looks at all my initial moves and finds the ones resulting in the best EV. It then looks at all my opponent's responses to each move and finds the ones resulting in the lowest EV for me (this now propagates up to become the EV of my initial move). It then looks at all my responses to my opponent's responses and finds the ones with the best EV (and does the propagation again). This is called min-max. Looking at all the best moves only is called alpha-beta search and is how most game programs work.

So the question is, what is "optimal"? On a scale of one to ten, if a move is three better than the next move, is that optimal? Let's say it is.

In chess, it's easy to see optimal moves. If someone does rook takes queen, then hopefully I can take his queen and it's a fair trade. By far that will be the only optimal move. Other optimal moves lead to checkmate or great increases in material. Otherwise, it's probably not optimal. In a typical chess game, maybe 5 percent of the moves have a value greater than "one pawn's worth."

In backgammon, I'd say it's 10 percent. I'm saying this based on experience with Backgammon NJ [an excellent program, BTW], discussions with backgammon game programmers in the past, and I'm using 10 percent rather than 5 percent because backgammon is slightly less complex than chess. It's not simple, though. To be a backgammon master probably requires almost as much study but not quite.

Hope this was helpful.

Yes, James, helpful indeed—because I now know a bit better how you think about the game, which I desperately need to finally beat you in our 101-point matches. Thanks!

WHAT ARE MY CHANCES OF MAKING THE CHAMPIONS TOUR (OR AT LEAST HITTING THE GOLF BALL REALLY FAR)? (SDL)

Despite the fact that I am not very good at golf, my secret fantasy is to someday play on the Champions Tour, the professional golf tour for fifty-somethings. As I approach my forty-fourth birthday, I realize it is time to get serious in this endeavor.

The right way to spend my time if I really wanted to make the tour, I suppose, would be to practice more. My friend Anders Ericsson popularized the magic number of ten thousand hours of practice to become an expert. Depending on exactly what you count as practice, by my rough calculations I have logged about five thousand hours of golf practice over the course of my life. Given how mediocre I am after the first five thousand hours, however, I'm not so optimistic that the next five thousand hours will lead me anywhere good.

So instead, I spent some time today figuring out how just how much I will need to improve. The best PGA Tour pros tend not to have regular handicaps, but are said to be the equivalent of Plus 8 on the handicap scale—i.e., eight strokes better than a scratch golfer. I claim to be a six handicap. That means that, to a first approximation, if I played eighteen holes today against the best players in the world, I should lose by fourteen strokes.

The probability that I will improve by fourteen strokes in the next six years is easy to estimate: zero.

Fortunately, my goal isn't to be the best golfer in the world, just to be the worst golfer on the Champions Tour. Surely, that can't be so hard, can it?

So I set out to measure just how much worse that guy is than the world's very best golfers. A direct comparison is hard to make because the bottom feeders on the Champions Tour rarely play against the Tiger Woodses of the world. The stars of the Champions Tour do, however, play an occasional PGA Tour event. I was able to find nineteen players who competed on both tours in 2010. On average, these players had a stroke average of 70.54 when playing on the Champions Tour, compared to an average of 71.77 when they played PGA Tour events. This suggests that the typical Champions Tour course plays a little more than one stroke easier than the typical PGA Tour course.

The top players on the PGA Tour post average scores of a little below seventy strokes per round, meaning that the upper echelon of senior golfers is about two strokes worse per round than the best players in the world. The low performers on the Champions Tour score around seventy-three on Champions Tour courses, or about two and a half strokes worse than the top senior golfers. If the world's best golfers are Plus 8 handicappers, then that means the "bad" golfers on the senior tour are roughly Plus 3 or Plus 4.

That's "only" nine or ten strokes a round better than me. Surely I can close that gap! If I can squeeze merely one stroke of improvement out of each incremental five hundred hours of practice, then by the time I hit ten thousand hours, I will be a Plus 4.

With that goal in mind, I recently started taking golf lessons for the first time since I was thirteen years old. One reason I chose my new golf coach, Pat Goss, is that he was an undergraduate economics major at Northwestern. I thought maybe he would understand the way I think.

On our first meeting, Pat first told me I swing like a character out of *Caddyshack*, and then asked me about my golf goals.

I responded with 100 percent honesty: "I want to play on the Champions Tour. But if you decide I'll never be that good, then I have a very different objective. I don't care the slightest bit about what my handicap ends up being in that case. All that matters to me then is being able to hit the ball as far as possible, even if I can't break a hundred."

I guess he's not used to getting an honest answer to this question, because he was so overcome with laughter he practically fell to the ground.

The good news is that six lessons later we are still devoting time to perfecting my short game, suggesting he thinks I can achieve my dream of making the tour.

Or maybe he's just maximizing revenue. After all, he is an economist by training.

10,000 HOURS LATER: THE PGA TOUR?
(SDL)

Last spring, I jokingly (okay, maybe half jokingly) wrote about my quest to make the Champions Tour, the professional golf tour for people over the age of fifty. In that post, I made reference to the ideas of Anders Ericsson, who argues that with ten thousand hours of the right kind of deliberate practice, more or less anyone can become more or less world class at any-thing. I've spent five thousand hours practicing golf, so if I could just find the time for five thousand more, I should be able to compete with the pros. Or at least that is what the theory says. My scorecards seem to be telling a different story!

It turns out I've got a kindred spirit in this pursuit, only this guy is dead serious. A few years back, twenty-something Dan McLaughlin decided he wanted to play on the PGA Tour. Never mind that he had only played golf once or twice in his life and had done quite poorly those times. He knew the 10,000-hour argument, and he thought it would be fun to give it a test. So he quit his job, found a golf coach, and has devoted his life to golf ever since. So far he is 2,500 hours into his 10,000-hour quest, which he chronicles at thedan plan.com.*

A while back I happened to find myself at Bandon Dunes, the golfers' haven on the Oregon coast. I met Dan there, and we had the chance to play thirty-six holes together. We had a great time, and it was fasci-nating to get to know him and hear about his approach.

The golf pro who has been guiding him had a very unusual plan, to say the least. For the first six months of Dan's golfing life, he was only allowed to putt. We are literally talking about Dan standing on a putting green for six to eight hours a day, six or seven days a week, hitting one putt after another. That is nearly one thousand hours of putting before he ever touched another club. Then he was given a wedge. He used just the wedge and the putter for another few months, before he got an eight iron. It wasn't until a year and a half into his golfing life—two thousand hours of practice—that he hit a driver for the first time.

I understand the basic logic of starting close to the hole (most shots in golf, after all, do occur close to the hole), but to my economist's mind, this sounds like a very bad strategy for at least two reasons.

First, one of the most basic tenets of economics is what we call diminishing marginal returns. The first little bit of something yields big returns; the more you do of something, the less valuable it is. For example, the first ice cream cone is delicious. The fourth is nauseating. The same must be true of putting. The first half hour is fun and engaging. By the eighth straight hour, it must be mind-numbing. I just can't imagine a person could focus that single-mindedly on putting, not just one day, but for months and months on end.

Second, my own experience suggests that there are spillovers across different aspects of golf. Things you feel when chipping help inform the full swing. Sometimes I can feel what I should be doing with a driver, and that helps me with my irons. Sometimes it is the opposite. To be putting and chipping for months without any idea what a full swing is—that just seems wrong to me.

So is the strategy working? After 2,500 hours, Dan is still really excited about golf, so that is a victory in and of itself. He is an eleven handicap, which means he is about fifteen to sixteen strokes per round away from being good enough for the PGA Tour. That means he has to shave off about one stroke for every five hundred hours of practice from here on out. I suspect he can keep that rate of improvement for the next few thousand hours, but it will be a tough haul after that.

Whatever the outcome, I'll be rooting for him. Partly because he is a nice guy, and partly because he promised me free tickets to the 2016 U.S. Open, but only if he qualifies.

*As of this writing (January 2015), Dan has just over 4,200 hours to go, and his handicap is down to 3.1.

LEVITT IS READY FOR THE SENIOR TOUR (SJD)

Levitt has made no secret of his desire to become a good enough golfer to someday play the Champions Tour, for players fifty and older.

After watching his amazing performance last week, I now believe Levitt does stand a chance of landing on the senior professional tour. But not in golf.

I was out in Chicago for a couple of days to work with Levitt. After a long day, we went out for dinner at a place near the University of Chicago called Seven Ten. It has food, beer, and bowling alleys—just a couple of them and nothing fancy. Old-school bowling.

After the meal, I tried to get Levitt to bowl a game or two. He wasn't interested. Said he was worried about hurting his golf swing. (Puh-leeze.) He said he'd watch me bowl. I can't think of anything less fun than bowling alone except having someone sit and watch you bowl alone. So I lied and told him that bowling would probably be good for his golf swing—the heavy ball could loosen up his joints, yada yada, etc.

He finally agreed when I suggested the loser pay for dinner.

He somehow found a ten-pound ball that fit his fingers, and in his first practice frame he rolled it as if it were a duckpin ball. It missed everything. I was feeling pretty good about the bet. Out of friendship, I suggested he try a heavier ball. He moved up to a twelve-pounder. And then he proceeded to bowl a 158, which he told me was about thirty pins above his average. He won.

There was nothing impressive about his form: even though he's a righty, he delivered from left to right and he put no movement on the ball. But he knocked the pins down.

So of course I suggested we bowl a second game. He said he wasn't interested but, again, he came around.

He opened with a spare and then a turkey—three strikes in a row. Amazing! Then two open frames. His luck had seemingly run out. But it hadn't: he now rolled four more consecutive strikes. It is hard to describe how unlikely this seemed, and was. He wound up with a 222. A 222! I took bowling as a PE requirement in college, and my career high is only 184.

When we got back to his house, Levitt looked up the current top PBA bowlers: a 222 average would put you firmly in the top twenty. And he bowled his 222 cross-lane, with a twelve-pound ball, after a big dinner, a beer, and a day of work.

My best explanation is that Levitt's maniacal devotion to golf, especially his thousands of hours of short-game practice, may have unwittingly turned him into a bowling dynamo. Either that, or he was lying about his existing average and he simply sandbagged me into buying dinner.

In either case, it was a pretty impressive feat. Unfortunately, his appearance on the PBA's senior tour is unlikely: wanting to go out on top, he has vowed never to bowl again.

True to his word, Levitt hasn't touched a bowling ball since.

LOSS AVERSION IN THE NFL
(SJD)

Football coaches are known for being extraordinarily conservative when it comes to calling risky plays, since a single bad decision (or even a good decision that doesn't work out) can get you fired. In the jargon of behavioral economics, coaches are "loss-averse"; this concept, pioneered by Amos Tversky and Daniel Kahneman, holds that we experience more pain with a loss of x than we experience pleasure with a gain of x. Who experiences loss aversion? Well, just about everyone: day traders, capuchin monkeys, and especially football coaches.

Which is why the last play of yesterday's Chiefs-Raiders game was so interesting. With five seconds left, Chiefs coach Dick Vermeil had a tough decision to make. His team was trailing by three points with the ball inside the Raiders' one-yard line. If the Chiefs ran a play and didn't score, they would likely not have time for another play and would lose. If they kicked the easy field goal, the game would go to overtime—and even though the Chiefs were playing at home, the Raiders had moved the ball easily late in the game, and Vermeil, as he would later admit, was scared that the Raiders would win the coin toss in overtime and promptly score, winning the game without the Chiefs ever having a chance.

In retrospect, it wasn't all that tough of a gamble. Choosing between a) a very significant gain if his team could accomplish the relatively simple act of advancing the ball two feet; or b) a shadowy outcome that seemed as likely to end in loss as in victory, Vermeil did what most of us would probably do if we didn't have several million people peeping over our shoulders, ready to criticize us: he went for the touchdown.

Vermeil sent in a running play, Larry Johnson dived into the end zone, and the Chiefs won. The front-page headline on today's *USA Today:* "Chiefs' Bold Gamble Hits Pay Dirt at Home: Kansas City shocks Oakland with touchdown after forgoing tying field goal on last play of game."

The fact that Vermeil's decision became the lead story is one good indicator of how rare it is for coaches to take such a risk. Here is what he later told reporters: "Wow! I was scared. I just figured I'm too old to wait. [Vermeil recently turned sixty-nine.] If we had not made it, then you guys would have had a lot of fun with that. It was not an impulsive thing. It was the right thing for us to do."

Congratulations to Vermeil for making a good choice that turned out well. Here's hoping a few of his colleagues will be envious enough of the attention he gets for making this wise gamble and follow suit.

BILL BELICHICK IS GREAT
(SDL)

I respect Bill Belichick more today than I ever have.

Last night he made a decision in the final minutes that led his New England Patriots to defeat. It will likely go down as one of the most criticized decisions any coach has ever made. With his team leading by six points and just over two minutes left in the game, he elected to go for it on fourth down on his own side of the field. His offense failed to get the first down, and the Indianapolis Colts promptly drove for a touchdown.

He has been excoriated for the choice. Everyone seems to agree it was a terrible blunder.

Here is why I respect Belichick so much. The data suggest that he actually probably did the right thing if his objective was to win the game. The economist David Romer studied years' worth of data and found that, contrary to conventional wisdom, teams seem to punt way too much. Going for a first down on fourth and short yardage in your own territory is likely to increase the chance your team wins (albeit slightly). But Belichick had to know that if it failed, he would be subjected to endless criticism.

If his team had gotten the first down and the Patriots won, he would have gotten far less credit than he got blame for failing. This introduces what economists call a "principal-agent problem." Even though going for it increases his team's chance of winning, a coach who cares about his reputation will want to do the wrong thing. He will punt just because he doesn't want to be the goat. (I've seen the same thing in my research on penalty kicks in soccer; kicking it right down the middle is the best strategy, but it is so embarrassing when it fails that players don't do it often enough.) What Belichick proved by going for it last night is that 1) he understands the data; and 2) he cares more about winning than anything else.

HOW ADVANTAGEOUS IS HOME-FIELD ADVANTAGE? AND WHY?
(SJD)

Do home teams really have an advantage?

Absolutely. In their book *Scorecasting*, Toby

Moscowitz and Jon Wertheim helpfully compile the percentage of home games won by teams in all the major sports. Some data sets go back further than others (MLB figures are since 1903; NFL figures are "only" from 1966, and MLS since 2002), but they are all large enough to be conclusive:

League	Home Games Won
MLB	53.9%
NHL	55.7%
NFL	57.3%
NBA	60.5%
MLS	69.1%

So it's hard to argue against the home-field advantage. In fact Levitt once wrote an academic paper about the wisdom of betting (*shh!*) on home underdogs, which we wrote about further in the *Times*.

But why does that advantage exist? There are a lot of theories to consider, including:

- "Sleeping in your own bed" and "eating home cooking"

- Better familiarity with the home field/court

- Crowd support

Those all make sense, don't they? In *Scorecasting*,

Moscowitz and Wertheim compile data to test a variety of popular theories. You might be surprised (and maybe even disappointed) to read their conclusion:

> When athletes are at home, they don't seem to hit or pitch better in baseball . . . or pass better in football. The crowd doesn't appear to be helping the home team or harming the visitors. We checked "the vicissitudes of travel" off the list. And although scheduling bias against the road team explains some of the home-field advantage, particularly in college sports, it's irrelevant in many sports.

So if these popular explanations don't have much explanatory power for home-field advantage, what does?

In a word: the refs. Moscowitz and Wertheim found that home teams essentially get slightly preferential treatment from the officials, whether it's a called third strike in baseball or, in soccer, a foul that results in a penalty kick. (It's worth noting that a soccer referee has more latitude to influence a game's outcome than officials in other sports, which helps explain why the home-field advantage is greater in soccer, around the world, than in any other pro sport.)

Moscowitz and Wertheim also make clear,

however, an important nuance: official bias is quite likely involuntary.

What does this mean? It means that officials don't consciously decide to give the home team an advantage— but rather, being social creatures (and human beings) like the rest of us, they assimilate the emotion of the home crowd and, once in a while, make a call that makes a whole lot of close-by, noisy people very happy.

One of the most compelling (and cleverest) arguments in favor of this theory comes from a research paper by Thomas Dohmen about home-field advantage in the Bundesliga, the top soccer league in Germany.

Dohmen found that home-field advantage was smaller in stadiums that happened to have a running track surrounding the soccer pitch, and larger in stadiums without a track.

Why?

Apparently, when the crowd sits closer to the field, the officials are more susceptible to getting caught up in the home-crowd emotion. Or, as Dohmen puts it:

> The social atmosphere in the stadium leads referees into favoritism although being impartial is optimal for them to maximize their re-appointment probability.

So it looks like crowd support does matter—but not in the way you might have thought. Keep this in mind next time you're shouting your brains out at a football game. Just make sure you know who you're supposed to be shouting at.

TEN REASONS TO LIKE THE PITTSBURGH STEELERS
(SJD)

After the 9/11 terrorist attack on New York, a lot of people wrote or called to ask if my family and I were okay. Some of these people were casual acquaintances at best but, for many of them, I was the only person they knew who lived in New York. Their concern was extremely moving even if, at first, a bit surprising.

I've been reminded of this outpouring over the past two weeks, as I've fielded e-mails and calls from people congratulating me on the Pittsburgh Steelers' making it back to the Super Bowl, against the Cardinals. I figure that, once again, for many of these people I am the only Steelers fan they know.

I feel sheepish accepting congratulations for an accomplishment as weak as this—simply rooting for a team that happens to win a bunch of football games. Plainly I can claim no credit. While it is true that I have brought my young son, a devout fan, to Pittsburgh for a game in each of the past three seasons, the Steelers lost all three games! Considering that their overall home record during that period was 13–6, I am obviously no good-luck charm.

But with great fortune comes great responsibility, and so, in return for this great fortune, let me accept the responsibility of laying out a few reasons to like the Steelers. I am not trying to convert anyone here; I'm only dispensing some ammunition for the undecided.

1. While the Steelers are attempting to win their record sixth Super Bowl, they were for the first forty years of their existence almost incomparably bad. So whether you gravitate toward prolific winners or lovable losers, the Steelers can satisfy your needs. Back in the 1930s, they paid big money to sign the college star Byron "Whizzer" White. He played wonderfully but stayed only one season; he went on to a slightly more impressive career as a U.S. Supreme Court justice.

2. The Steelers have been majority-owned by the same family, the Rooneys, since the team's founding in 1933. The story goes that Art Rooney

bought the team for $2,500 with the winnings of a great day at Saratoga Race Course—he was a vigorous gambler and a beloved rogue—but that is probably apocryphal. The team is now onto its third generation of family management and, as families go, the Rooneys are pretty exemplary: honorable, charitable, humble, and more. (If you are pleased with Barack Obama, you have extra reason to like them. Dan Rooney, the team's seventy-six-year-old chairman, is a lifelong Republican who last year got behind Obama early and campaigned hard throughout Pennsylvania. It'd be a stretch to say that Rooney pushed the election toward Obama, but there are few brands in the state as strong as the Steelers, so it certainly didn't hurt.) The family prides itself on running a football team that reflects its values; the Steelers are known as a "character" team. Which makes it interesting to see what happens when a player exhibits some bad character. Earlier this season, when starting wide receiver Santonio Holmes was pulled over by the police for marijuana possession (he used to sell drugs as a teenager, it turns out), the team suspended him for a week. This was hardly mandatory; Holmes hadn't even been arrested. But it sends a signal.

Meanwhile, a starting wide receiver for the San Diego Chargers, Vincent Jackson, was arrested for suspicion of drunk driving a few days before the Chargers recently came to Pittsburgh for a playoff game. The Chargers issued one of those pro forma "we will monitor the situation" press releases, and Jackson played as usual.

3. Myron Cope. He was a talented writer who became a Steelers broadcaster despite having a voice that sounded like gravel and Yiddish tossed in a blender. He was relentlessly unique; among his on-air exhortations: "Yoi!" or, if something really exciting happened, "Double Yoi!" Cope, who died last year, deftly blended boosterism with realism, which made him an institution in Pittsburgh. But the accomplishment for which he'll remain best known is inventing the Terrible Towel, a Steelers-gold terry-cloth rag that will be widely seen, waving madly in the Tampa sunshine, on Sunday. Many other teams have copied the Towel, but nowhere does it have such resonance as in Pittsburgh—in part because Cope donated the considerable profits to the Allegheny Valley School, a home for people with intellectual and developmental disabilities, whose residents include Cope's son.

4. The fan diaspora. Even though the city of Pittsburgh has transformed itself nicely from a manufacturing town to a service town, it has lost about half its population in the last few decades. This has created a diaspora of fans all over the country and beyond, Steelers lovers who had to leave the 'Burgh for better jobs and who then taught their kids to be Steelers lovers even though they lived in Arizona or Florida or Alaska. As a result, there's a "Steelers bar"—a place to watch the game on Sundays with like-minded folks—in just about any good-sized city in America. The Steelers may not be "America's Team," as the Cowboys claim, but perhaps they should be.

5. Franco Harris. One of the most interesting and enigmatic football players in history, so much so that somebody (i.e., yours truly) even wrote a book about his strange appeal. Franco was also, of course, the star of the football miracle known as the Immaculate Reception (whose name was popularized, naturally, by Cope). Also, his teammate Mean Joe Greene was the star of one of the best TV commercials ever—which is being remade this year, with the extraordinarily appealing Troy Polamalu in the lead.

6. The Steelers are good assessors of talent, both seen and unseen. Consider their first-round draft picks since 2000: Plaxico Burress, Casey Hampton, Kendall Simmons, Troy Polamalu, Ben Roethlisberger, Heath Miller, Santonio Holmes, Lawrence Timmons, and Rashard Mendenhall. Aside from Burress, all but two are valuable Steelers starters. Timmons is on the verge of being a valuable starter and it's too early to tell about Mendenhall, the rookie whose shoulder was broken in mid-season by Ray Lewis. And, even more impressively, consider the fact that two of their very best players, Willie Parker and James Harrison, were undrafted. Harrison, recently named the league's defensive MVP, is the only undrafted player in history to have won this award. (Granted, the Steelers' opponents in the Super Bowl, the Arizona Cardinals, are quarterbacked by Kurt Warner, a potential Hall of Famer who was bagging groceries for a living before he made it as a football player.)

7. The Steelers are a small-market team (Pittsburgh's population is less than 350,000) that manages to always play big. Compare them

to Pittsburgh's baseball team, the Pirates, which hasn't had a winning season in fifteen years. True, small-market teams have an easier time in football than in baseball because of the NFL's revenue-sharing policy, but it's also true that the Steelers are a fiscally prudent organization. This can especially be seen in their willingness to let their own high-priced free agents go (Alan Faneca, Joey Porter, and Plaxico Burress are recent examples). Nor do they purchase the rights of aging superstars who won't fit their team anyway.

8. Especially when compared to baseball, there is a real paucity of great books about football. One of the very best, however, Roy Blount Jr.'s *About Three Bricks Shy of a Load,* is about the Steelers.

9. Mike Tomlin, the current head coach, is a young and impressive man brimming with smarts, balance, dignity, and surprise. (At the press conference immediately after the Steelers beat the Baltimore Ravens to gain the Super Bowl, he quoted Robert Frost.) Tomlin was hired just two years ago. The Steelers' previous two head coaches, Chuck Noll and Bill Cowher, lasted a

combined thirty-seven years. These days, NFL coaches are chewed up and spat out with abandon, often within two or three years, but I have a feeling Tomlin may end up threatening Cowher and Noll for longevity awards.

10. The Steelers are one of the few pro sports teams named after what their respective cities actually do or did. Pittsburgh made steel just as Green Bay packed meat; the cardinal, meanwhile, is a perfectly nice bird, but it doesn't do squat for Arizona (nor did it previously for St. Louis). Furthermore, the Steelers' logo isn't a cartoonish bird or patronizingly noble "redskin"; it is the actual mark of steelmaking—a trio of red, blue, and yellow hypocycloids in a black circle. Plus, the Steelers wear it only on one side of their helmets. Legend holds this is because the team was so frugal that it didn't want to use up two decals on each helmet.

You are free, of course, to ignore all of the above and root for the Cardinals (a team that happens to comprise a bunch of former Steelers coaches, players, and even a onetime ball boy). But if you do choose to cheer on the Steelers, know that there are some good reasons for doing so.

All the cheering may have done some good: the Steelers beat the Cardinals 27–23.

THE MAKING OF A FIRST-GRADE DATA HOUND
(SJD)

My son's first-grade teacher recently held an open house to tell the parents what their kids will be learning this year. I have to say, it was pretty impressive. My favorite part had to do with turning the kids into first-grade (if not first-rate) empiricists.

The teacher, a wonderful veteran from Texas named Barbara Lancaster, described an upcoming project: collecting data on some or all of the twenty-two playgrounds in Central Park.

First, the kids will vote on their favorite and least favorite playgrounds in the park. Then they will collect data on a variety of metrics: number of swings, amount of open space, shady vs. sunny areas, etc. Then they will try to figure out the factors that make a good playground good and a bad playground bad. They will also consider the safety of each playground, and other measures.

We did not do this kind of project when I was in first grade; frankly, I am envious.

I recently played a game with my kids in Central Park that is along these same lines. We sat on one of our favorite rocks overlooking the Loop, a six-mile road that runs through the park, and I asked if they thought there were more runners or cyclists going past. Both kids were certain that there were more cyclists—perhaps because, since the cyclists are so much faster than the runners, they make more of an impression. So we made a little bet (I took runners, they took cyclists), and started counting to see which would pass us first: a hundred runners or a hundred cyclists. I won, but not by much: 100–87.

That was early on a weekday evening. But a few days later, we played the same game on a weekend morning. The kids stuck to their guns and picked cyclists. This time they were right: the cyclists blew away the runners. I guess there are a lot of people who aren't willing to unpack their bikes for a weekday evening ride, especially as the days are getting shorter, but are willing to go to the trouble on a weekend morning.

It was a good lesson for all of us, and it has us on the lookout for other fun things to measure. The even better lesson is that it's probably a good idea to always combine teaching—whether yourself or your kids—with playing a game.

ANNUAL KENTUCKY DERBY PREDICTIONS
(SDL)

I'm not sure why, since I don't think anyone should or does care, but every year I indulge myself by posting Kentucky Derby picks.

In contrast to the last two years, my computer model has some strong predictions for this year's Derby. The two horses I like best from a betting perspective (i.e., the ones I think actually have a positive expected value if you place a win bet on them) are General Quarters and Papa Clem. Both are long shots, with morning-line odds of 20–1, but my model predicts their actual odds will be lower than this.

There are other horses who look pretty good, too, although not strong enough to yield a positive expected value on a bet: Friesan Fire, Musket Man, and Flying Private.

The favorite, I Want Revenge, looks okay also, but not good enough to bet.

If I had to pick a last-place finisher (a bet they would never actually offer at the track because people involved with horse racing understand better than most that people respond to incentives), it would be Mine That Bird.

A FEW DAYS LATER . . .

THANKFULLY, NO ONE PAYS ATTENTION
(SDL)

Thankfully, no one pays attention to my annual Kentucky Derby picks, because if they did, they would have read this prediction that I made Friday:

"If I had to pick a last-place finisher (a bet they would never actually offer at the track because people involved with horse racing understand better than most that people respond to incentives), it would be Mine That Bird."

And then they would read this headline in *The Boston Globe*'s Sunday sports page:

50-1 SHOT SHOCKS FAVORITES AT DERBY
Mine That Bird runs away with 6 3/4-length victory

But it gets worse before it gets better. I named five horses that I did like. One of these horses finished dead last out of eighteen horses, another finished next to last!

The other three finished respectably: third, fourth, and tenth.

In light of this showing, will there be any demand for my Preakness picks? I think so. You see, when it comes to predictions, there is as much value in someone who can predict as poorly as I do as there is in someone who predicts the truth. You just take the bad predictor's picks and do the opposite.

Chapter 9
When to Rob a Bank

 *We have always been attracted to crime—not doing it, necessarily, but exploring it. One of the most-talked-about subjects in Freakonomics was our argument that the legalization of abortion was responsible for driving down crime twenty years later. Our love of crime almost landed us a network TV show, almost got one of us sent to Guantánamo Bay, and did inspire the title of this book.*

WHEN TO ROB A BANK

(SJD)

I recently read about a man who robbed six banks in New Jersey but only on Thursdays. "No reason was given for choosing that particular day," the article noted.

Perhaps he knew something about how the banks did business; perhaps his astrologist told him Thursdays were lucky; perhaps it simply fit his schedule.

In any case, it reminded me of a story I heard upon a recent visit to Iowa, about a local bank employee named Bernice Geiger. She was arrested in 1961 for embezzling more than $2 million over many years. The bank happened to be owned by her father. Bernice was reportedly very generous, giving away lots of the stolen money. Upon her arrest, the bank went bust. She was sent to prison, paroled after five years, and moved back in with her parents, who were apparently forgiving types.

By the time she was arrested, Geiger was said to be exhausted. Why? Because she never took vacations. This turned out to be a key component in her crime. As the story goes—this was told to me by a retired Sioux City cop, though I haven't been able to confirm it—the reason she never took vacations was that she was keeping two sets of books and couldn't risk a fill-in employee discovering her embezzlement.

The most interesting part, according to the cop, is that after prison Geiger went to work for a banking oversight agency to help stop embezzlement. Her biggest contribution: looking for employees who failed to take vacation. This simple metric turned out to have strong predictive power in stopping embezzlement. Like cheating schoolteachers or colluding sumo wrestlers, the people who steal money from banks sometimes leave telltale patterns—whether it's a lack of vacation or a string of Thursdays—that point the finger right at them.

All this made me curious about bank-robbery statistics in general. Maybe Thursday really is the best day to rob a bank?

According to the FBI, there are roughly 5,000 bank robberies a year in the U.S. Friday is easily the busiest weekday (there are relatively few robberies on the weekend), with 1,042 Friday robberies in a year; this is followed by Tuesday (922), Thursday (885), Monday (858), and Wednesday (842). But there is no evidence that any one day is more likely to be successful than another.

It also appears as though robbers are not very good at maximizing their return. Morning robberies yield far more money than afternoon robberies ($5,180 versus $3,705) and yet bank robbers are far more prone to strike in the afternoon. (Maybe they like to sleep in? And maybe if they were able to wake up earlier and go to work, they wouldn't have to rob banks?) Overall, U.S. bank robbers earn an average of $4,120 when they are successful. But they aren't successful as often as I would have thought: they are arrested 35 percent of the time! So the Jersey robber who made it to his sixth Thursday was ahead of the pack.

The success rate of bank robbers in the U.K. is about the same as in the U.S., but British robbers generally get a lot more money. The economists Barry Reilly, Neil Rickman, and Robert Witt got hold of a trove of robbery data from the British Bankers' Association and analyzed it for a paper in *Significance*, a publication of the Royal Statistical Society. They found that the average proceeds from all bank raids, including the unsuccessful ones, was roughly $30,000. Robberies with multiple robbers, they note, tend to yield considerably bigger jackpots. Overall, the average raid yields about $18,000 per robber. So that's substantially higher than their American counterparts. But again, the likelihood of arrest is high. This leads the authors to conclude that "the return on an average bank robbery is, frankly, rubbish" and that "as a profitable occupation, bank robbery leaves a lot to be desired."

So if we want to know when is the best time to rob a bank, the answer would seem to be . . . never. Unless, of course, you happen to work at one. But even then, the trade-off is steep—for you may have to give up vacations forever.

WHAT'S THE REAL CRIME RATE IN CHINA?

(SDL)

Official statistics would certainly suggest that crime in China is extremely low. Murder rates are roughly one-fifth as high as in the United States; according to the official crime statistics, all crimes are rare. China certainly feels safe. We walked the streets in rich areas and poor areas and not for a moment did I ever feel threatened. Graffiti was completely absent. The one instance where I thought I finally found some graffiti near a train station in the city of Shangrao, the spray-painted message on a bridge turned out to be a government warning that anyone caught defecating under the bridge would be severely punished.

Yet there were all sorts of odd behaviors that made it seem like some crimes were a big problem.

First, there seemed to be an obsession with the risk of counterfeit money. Our tour guides felt the need to teach us how to identify fake money. Whenever I bought something with currency, the shopkeeper went through a variety of tricks to validate the legitimacy of the bills.

Second, when checking out of some of our hotels, there was a fifteen-minute delay while a hotel worker went to inspect the vacated hotel room, I presume to look for stolen clocks, towels, and mini-bar items. (Possibly I misunderstood why they checked the room, just as I struggled to understand why there was a fifteen-dollar fee associated with each lost key card, which couldn't possibly have cost the hotel more than a few cents.)

Third, places that no sensible person would ever want to break into (for instance, orphanages) were protected by guardhouses and metal gates that had to be retracted to let vehicles in. I don't think the gates were to keep the orphans in, but maybe they were!

Fourth, on the trains we took, they checked our tickets before we boarded, while we were riding the train, and also required that the ticket be produced on the way out of the station.

Finally, and most notably, public restrooms were completely devoid of toilet paper, even in some reasonably nice restaurants. Again, maybe I'm completely missing something, but the impression was that a) toilet paper was a very valuable commodity; and b) if it were left in public restrooms it would be stolen.

DON'T REMIND CRIMINALS THEY ARE CRIMINALS
(SDL)

Psychologists have long argued about the power of priming—i.e., the power of subtle cues and reminders to influence behavior. For instance, there are a number of academic papers that find that if you make a woman write down her name and circle her gender before taking a math test, she will do substantially worse than if she just writes her name. The idea is that women perceive that they are not good at math, and circling their gender reminds them that they are women and therefore should be bad at math. I've always been skeptical of these results (and indeed failed to replicate them in one study I did with Roland Fryer and John List) because gender is such a powerful part of our identities that it's hard for me to believe that we need to remind women that they are women!

In an interesting new study called "Bad Boys: The Effect of Criminal Identity on Dishonesty," Alain Cohn, Michel André Maréchal, and Thomas Noll find some fascinating priming effects. They went into a maximum security prison and had prisoners privately flip coins and then report how many times the coin came up "heads." The more "heads" they got, the more money they received. While the authors can't tell if any one prisoner is honest or not, they know that on average "heads" comes up half the time, so they can measure in aggregate how much lying there is. Before the study, they had half the prisoners answer the question "What were you convicted for?" and the other half

"How many hours per week do you watch television on average?" The result: 66 percent "heads" in the treatment where they ask about convictions and "only" 60 percent "heads" in the TV treatment.

How dishonest are prisoners versus everyday people? When they play the same game with regular citizens, the coin comes up "heads" 56 percent of the time.

So how powerful is that one question on conviction? The behavior of the prisoners who are asked the TV question is actually closer to that of regular citizens than it is to the behavior of the prisoners who were primed.

As an economist, I hate the idea that priming might work. As an empiricist, I guess I better get used to it.

WHAT DO REAL THUGS THINK OF *THE WIRE*? (SUDHIR VENKATESH)

Sudhir Venkatesh has become a familiar figure to Freakonomics readers. During graduate school in Chicago, he embedded himself for several years with a crack gang; this research formed the backbone of our book chapter called "Why Do Drug Dealers Still Live with Their Moms?" He continued to do fascinating research at the lowest and highest ends of the economic spectrum, often writing about it for the Freakonomics blog.

Ever since I began watching HBO's *The Wire*, I felt that the show was fairly authentic in terms of its portrayal of modern urban life—not just the world of gangs and drugs, but the connections between gangland and City Hall, the police, the unions, and practically everything else. It certainly accorded with my own fieldwork in Chicago and New York. And it was much better than most academic and journalistic reportage in showing how the inner city weaves into the social fabric of a city.

A few weeks ago, I called a few respected street figures in the New York metro region and asked them to watch the show's new season. I couldn't think of a better way to ensure quality control.

For the first episode, we gathered in the Harlem apartment of Shine, a forty-three-year-old half-Dominican, half–African-American man who managed a gang for fifteen years before heading to prison for a ten-year drug-trafficking sentence. I invited older guys like Shine, most of whom had retired from the drug trade, because they would have greater experience with rogue cops, political toughs, and everyone else that makes *The Wire* so appealing. They affectionately named our gathering "Thugs and Cuz." ("Cuz"—short for "cousin"—was me.)

There was plenty of popcorn, ribs, bad domestic beer, and fried pork rinds with hot sauce. The pork rinds, apparently the favorite of the American thug, ran out so quickly that one of the low-ranking gang members in attendance was dispatched to acquire several more bags.

Here's a quick-and-dirty summary of the evening's highlights:

1. The Bunk is on the take. Much to my chagrin (since he is my favorite character), the consensus in the room was that the Bunk was guilty. In the words of Shine, "He's too good not to be profiting. I got nothing against him! But he's definitely in bed with these street [thugs]." Many had known of Bunk's prowess as a detective from past episodes. The opening scene, in which he craftily obtains a confession, reinforced their view that the Bunk is too good not to be hiding something.

2. Prediction number one: McNulty and the Bunk will split. The observation regarding Bunk's detective work led to a second agreement, namely

that McNulty or Bunk will be taken down—shot, arrested, or killed. This was closely tied to the view that McNulty and Bunk will come into conflict. The rationale? Everyone felt that Marlo, Proposition Joe, or another high-ranking gang leader must have close (hitherto unexplained) ties with one of these two detectives. "Otherwise," said Kool-J, an ex–drug supplier from northern New Jersey, "there ain't no way they could be meeting in a Holiday Inn!" Orlando, a Brooklyn based ex–gang leader, believed the ambitions of Bunk and McNulty would run into each other. "One of them will be taken down. Either the white boy gets drunk and shoots some [guy] 'cause he's so pissed, or Bunk gives him up to solve a case!"

3. The greatest uproar occurred when the upstart Marlo challenged the veteran Prop Joe in the co-op meeting. "If Prop Joe had balls, he'd be dead in twenty-four hours!" Orlando shouted. "But white folks [who write the series] always love to keep these uppity [characters] alive. No way he'd survive in East New York more than a minute!" A series of bets then took place. All told, roughly $8,000 was wagered on the timing

of Marlo's death. The bettors asked me—as the neutral party—to hold the money. I delicately replied that my piggy bank was already full.

4. Carcetti is a fool. Numerous observers commented on the Baltimore mayor's lack of "juice" and experience when it came to working with the feds. The federal police, in their opinion, love to come in and disrupt local police investigations by invoking the federal racketeering statutes ("RICO") as a means of breaking up drug-trafficking rings. "When feds bring in RICO, local guys feel like they got no [power]," Tony-T explained, offering some empathy to local police who get neutered during federal busts. "White boy [a.k.a. Carcetti], if he knew what he was doing, would keep them cops on Marlo just long enough to build a case—then he would trade it to the feds to get what he wanted." Others chimed in, saying that the writers either didn't understand this basic fact, or they wanted to portray Carcetti as ignorant.

The evening ended with a series of additional wagers: Tony-T accepted challenges to his claim that Bunk dies by the end of the season; Shine proposed that Marlo would kill Prop Joe; the youngest attendee, the twenty-nine-year-old Flavor, placed $2,500 on Clay Davis escaping indictment and revealing his close ties with Marlo.

I felt obliged to chime in: I wagered five dollars that the circulation of *The Baltimore Sun* would double, attracting a takeover by Warren Buffett by Episode 4. No one was interested enough to take my bet.

Venkatesh went on to write nine columns about watching The Wire *with his criminally inclined friends. They can all be found at Freakonomics.com.*

THE GANG TAX
(SUDHIR VENKATESH)

New York's state senate recently passed a bill making it illegal to recruit someone into a street gang.

In the never-ending fight by city officials and legislators to combat gangs, this is one of the latest efforts to outmaneuver them. Other initiatives have included: city ordinances that limit two or more gang members from hanging out in public space; school codes that ban the use of hats, clothing, and colors that signify gang membership; and public housing authorities that evict leaseholders who allow gang members (or any other "criminal") to live inside the housing unit.

These laws rarely lead to reductions in gang membership, gang violence, or gang crime. In fact, police officers I know find these ordinances and statutes a waste of time. Cops would much rather "control and contain" gang activity. Most officers who work in inner cities understand that you cannot eliminate gang activity entirely—arrest two gang members and you will find a dozen others waiting in line to take their places. Police know that gang members have great knowledge about local crimes, so they rely on a trade-off: keep gangs isolated to particular areas, don't let their criminal activities spill over into other spaces, and use high-ranking gang members for information.

This strategy actually prevents membership from expanding, at least in big cities where gangs are economically oriented. Beat cops who run the streets make sure that gang leaders don't prey upon too many kids for recruiting purposes. In effect, this kind of policing limits the reach of gangs. It may not be socially desirable policing, but it works if you measure effectiveness by reductions in gang membership.

I called a few gang leaders in Chicago and asked them about the greatest obstacles to recruiting and retaining members. Here are a few answers:

Michael (thirty years old, African American) was insistent that today's gangs are mostly "drug crews," i.e., businesses:

We always lose people to jobs. If a n----r in my crew gets a good job, he's gone. So, as long as there ain't no work for a brother, then we have no problem. Most of us have families, we're not in school beating each other up, acting stupid. We're out here on the streets trying to make our money. You got all these people telling us to get an education—I'm making thousands of dollars each month. Why do I need to go to school?

Darnell (thirty-two years old, African American) said police should be more creative.

Let's say you catch one of us—I'd make the boy wear a dress and makeup. Maybe for two weeks. Let the boy go to school looking like a girl. Let him walk the streets looking like he's gay. I guarantee you, we'd have a hard time holding on to n----rs if you do shit like that!

Jo-Jo (forty-nine years old, half Puerto Rican, half black) said the cops should do . . .

. . . what they did when I was younger. Drop a Disciple off in Vicelords territory late at night. Let him get his ass kicked. And keep doing it! I remember growing up and all these cats used to get beat up. You know what? This would actually help me because it would get rid of a lot of these folks who do nothing for us except cause trouble. In fact, I'd be willing to work with the cops if they want to call me. Maybe we could help each other out?

My good friend Dorothy never ran a gang, but as an outreach worker who helps young people in the ghetto turn their lives around, she has pretty good insight. She recalled some of her own gang-intervention efforts in the 1990s and came up with the following suggestion:

Tax the n-----s! That's what I would do if I was the mayor. Don't put them in jail, but take fifty percent of their money. You know what I mean? Find them on the streets if they're misbehaving, grab half of their cash, and put it into a community fund. Let the block clubs have it, let the churches have the money. I guarantee you that a lot of brothers will think twice if you get to their pocketbooks.

Interesting thought. I wonder whether market forces might exert the kind of discipline required to limit the involvement of young people in gang-controlled drug economies. If, as Treasury Secretary Paulson reminds us, "market discipline" is sufficient to regulate the financial markets, perhaps it could be effective in the underground.

Oh, yes, I forgot about Bear Stearns. (Sorry, couldn't resist.)

DON'T BURN THE FOOD
(SDL)

In a sample of thirteen African countries between 1999 and 2004, 52 percent of women surveyed say they think that wife-beating is justified if she neglects the children; around 45 percent think it's justified if she goes out without telling the husband or argues with him; 36 percent if she refuses sex, and 30 percent if she burns the food.

And this is what the women think.

We live in a strange world.

WHEN WAS THE LAST TIME SOMEONE ANSWERED "YES" TO ONE OF THESE QUESTIONS?
(SDL)

In order to become a U.S. citizen, one has to complete the Immigration and Naturalization Service's Form N-400.

How long do you think it has been since someone answered yes to question 12(c) in part 10(b):

Between March 23, 1933, and May 8, 1945, did you work for or associate in any way (either directly or indirectly) with any German, Nazi, or SS military unit, paramilitary unit, self-defense unit, vigilante unit, citizen unit, police unit, government agency or office, extermination camp, concentration camp, prisoner of war camp, prison, labor camp, or transit camp?

I also wonder what kind of person answers yes to this question:

Have you ever been a member of or in any way associated (either directly or indirectly) with a terrorist organization?

I'm surprised we still bother to ask this question:

Have you ever been a member of or in any way associated (either directly or indirectly) with the Communist Party?

There are some trickier questions, though, like this one:

Have you ever committed a crime or offense for which you were not arrested?

Not many people can truthfully answer no to that last question, but I presume everyone does anyway.

Is there any point to asking questions when you know that people will never give a yes answer?

It turns out that there actually is a point to such questions: U.S. law enforcement can use demonstrably false answers against individuals to prosecute or deport them. Indeed, some officers I was speaking with the other day said they wished there were more questions on terrorist activities on the N-400.

IS PLAXICO BURRESS AN ANOMALY?
(SJD)

A few years back, I wrote an article for the *Times Magazine* about the NFL's annual "rookie symposium," a four-day gathering during which the league tries to warn incoming players about all the pitfalls they may face—personal threats, bad influences, gold-digging women, dishonest money managers, etc.

The NFL even brought in a bunch of veterans and retirees to try to teach the young guys some lessons. One was the former wide receiver Irving Fryar. As I wrote:

"We're going to have some idiots come out of this room," he begins. "Those of you feeling good about yourselves, stop it. You ain't did nothing yet." Fryar recites his career stats: 17 NFL seasons, a drug habit since he was 13, and four trips to jail. "The first time, I was stopped in New Jersey," he says. "I was on my way to shoot somebody. Driving my BMW. I had guns in the trunk, and I got taken to jail. The second time, also guns. Third time was domestic abuse. Fourth time, it was guns again. No. Yeah, yeah, it was guns again. Things got so bad for me, I put a .44 Magnum up to my head and pulled the trigger." Now Fryar is a minister. "When I was a rookie," he says, "we didn't have anything like this [symposium]. I had to learn it the hard way. Don't use me as an example of what you can get away with, brothers. Use me as an example of what you shouldn't do."

It looks like Plaxico Burress didn't pay attention. I briefly met Burress back when he was first drafted by the Pittsburgh Steelers, and have followed his career medium closely ever since. I have concluded that my first impression was pretty much accurate: he is a grade-A knucklehead. His latest misstep—shooting himself in the leg in a nightclub—is easily the most serious (he may well go to prison for criminal possession of a handgun under New York City law), but his history on and off the field reads like an idiot's checklist.

But how anomalous is Burress? According to an ESPN report, not very. One insider estimated that 20 percent of Major League Baseball players carry concealed weapons. A former cop who has worked as a bodyguard for NBA players put the number as "close to 60 percent." As for the NFL? Here's what ESPN reports: "New England Patriots wide receiver Jabar Gaffney, a gun owner himself, said he thinks 90 percent of NFL players have firearms."

Burress's problem—aside from the fact that he shot himself—is that he didn't have a carry permit. And while he lives in New Jersey, the shooting took place in New York City, where Mayor Michael Bloomberg is devoutly anti-gun.

If the ESPN figures are even halfway true, the question arises: Is the risk of carrying an illegal handgun smaller than the risk that the average NFL player faces if he goes out in public without a gun?

Burress would seem to have thought so.

Of all the stories about players who've gotten into trouble with guns, there's also the case of Sean Taylor, who was shot to death in his own home even though he was armed and tried to defend himself.

His weapon? A machete.

FORGET ABOUT HAVING YOUR FRIENDS OVER FOR DINNER; IN MISSOURI IT'S YOUR ENEMIES YOU WANT TO INVITE
(SDL)

For years, I've fantasized about buying a gun. The only reason I want one is that if an intruder enters my house and tries to terrorize my family, I would like to be able to defend us. The baseball bat under the bed just doesn't seem sufficient. Never mind that I am a total coward—at least I'd be able to imagine the scenario would play out differently.

Given my own heroic fantasies, I heartily endorse a new law passed in Missouri which stipulates that you can use deadly force against someone who illegally enters your home (or even your car), even if you aren't in obvious danger. In most places, you need to prove you were in real danger of being hurt or killed in order to justify the use of deadly force.

From a crime deterrence theoretical perspective, this law makes sense to me. A burglar has no legitimate reason to be in your house. Burglary is a crime with high social costs (victims feel an awful sense of violation when their home is ransacked, even if the burglar doesn't get much), but relatively low expected punishments for the criminal because arrest rates are low. Most victims never see the burglars, so they're difficult to catch, as opposed to street robberies. I did a rough calculation many years ago, and if I remember correctly, the risk of lost years of life for burglars who were shot and killed by their victim amounted to about 15 percent of the total prison time they could expect to serve for their crimes. In other words, if you are a burglar, being killed by the resident should be a serious concern. If this law encouraged more residents to kill intruders, there would likely be fewer burglaries.

On the other hand, this law probably won't have much real impact on crime. The kind of people that shoot burglars when they catch them in their homes are likely to shoot the burglar whether such a protective law is in place or not. (That is, more or less, my reading of the evidence on concealed-weapons laws.) I think that, in practice, they mostly let you off the hook legally if you shoot an intruder. If victim behavior doesn't actually change, there is little reason for burglar behavior to shift. Even worse, you get a bunch of bumblers like me trying to fight burglars under the new law, and we end up getting shot.

The law does bring to mind some interesting possibilities, however. If there is someone you dislike so much that you want him dead, all you need to do is figure out how to get him to come inside your house, and make it look plausible that he was an intruder. Maybe you could tell him that you are having a late-night poker party and to just let himself in and come upstairs to join the game. Or maybe say there's a surprise party for a mutual acquaintance, so all the lights will be out, and to come to your bedroom at 2 A.M.

Never underestimate the creativity and deviousness of humans—or the speed with which *Law and Order* will take the first example of this and turn it into an episode.

NO MORE D.C. GUN BAN? NO BIG DEAL (SDL)

The Supreme Court recently struck down the gun ban in Washington, D.C. A similar gun ban in Chicago may be the next to go.

The primary rationale for these gun bans is to lower crime. Do they actually work? There is remarkably little academic research that directly answers this question, but there is some indirect evidence.

Let's start with the direct evidence. There have been a few academic papers that directly analyzed the D.C. gun ban, and the papers came to opposite conclusions.

The fundamental difficulty with this kind of research is that you have one law change. So you can compare D.C. before and after. Or you can try to find a control group and compare D.C. before and after to that control group before and after (in what economists call a "differences-in-differences analysis").

The problem here is that crime rates are volatile and it really matters what control group you pick. I would argue that the most sensible control groups are other large, crime-ridden cities like Baltimore or St. Louis. When you use those cities as controls, the gun ban doesn't seem to work.

What about indirect evidence? In Chicago we have a gun ban and 80 percent of homicides are done with guns. The best I could find about the share of homicides done with guns in D.C. is from a blog post which claims 80 percent in D.C. as well. Nationwide that number is 67.9 percent, according to the FBI.

Based on those numbers, it is hard for someone to argue with a straight face that the gun ban is doing its job. (And it is not that D.C. and Chicago have unusually low overall homicide rates either.)

It seems to me that these citywide gun bans are as ineffective as many other gun policies are for reducing gun crime. It is extremely difficult to legislate or regulate guns when there is an active black market and a huge stock of existing guns. When the people who value guns the most are the ones who use them in the drug trade, there is next to nothing you can do to keep the guns out of their hands.

My view is that we should not be making policies about gun ownership, because they simply don't work. What seems to work is harshly punishing people who use guns illegally.

For instance, if you commit a felony with a gun, you get a mandatory five-year add-on to your prison sentence. Where this has been done there is some evidence that gun violence has declined (albeit with some substitution toward crimes being done with other weapons).

These sorts of laws are attractive for many reasons. First, unlike other gun policies, they work. Second, they don't impose a cost on law-abiding folks who want to have guns.

WHAT'S THE BEST WAY TO CUT GUN DEATHS?
(SJD)

Are there more guns in the U.S. or more opinions about guns?

Hard to say. We have written widely about guns over the years. Here we present a quorum with a narrow focus: What are some good ideas to cut gun deaths? Let's put aside momentarily the standard discussion about the right to bear arms and deal instead with the reality on the ground: there are a lot of gun deaths in this country; how can they be lessened?

We asked a few people who think about this issue a simple question: What's your best idea to cut gun homicides in the U.S.? You may not personally like these answers, but it strikes me that most of them are more sensible than what you typically hear in the gun debate these days.

Jens Ludwig is the McCormick Foundation professor of social service administration, law, and public policy at the University of Chicago's Harris School.

We should give out rewards—I mean big, serious rewards—for tips that help police confiscate illegal guns.

More people die from gun suicides than homicides in the U.S., but gun crime accounts for most of the $100 billion in social costs that Phil Cook and I estimate gun violence imposes each year. Most murders are committed with guns (around 75 percent in 2005 in Chicago). We also know that young people—particularly young males—are vastly over-represented among offenders; most murders happen outdoors; and a large share of all homicides stem from arguments or something related to gangs. A big part of America's problem with gun violence stems from young guys walking or driving around with guns and then doing stupid things with them.

Young guys carry guns in part because this helps them get some street cred. For a project that Phil Cook, Anthony Braga, and I conducted with the sociologist Sudhir Venkatesh (published in *Economic Journal*), Venkatesh asked people on the South Side of Chicago why they carry guns. As one gang member said, in the absence of having a gun:

"Who [is] going to fear me? Who [is] going to take me seriously? Nobody. I'm a pussy unless I got my gun."

Guns are something that a lot of guys seem to have mostly to take to football and basketball games or parties and to show off to their friends or girlfriends. At the same time, the costs of carrying guns might be low. A previous Freakonomics post by Venkatesh notes that cops are less likely to be lenient for other offenses if someone is caught with a gun. But the chances of being arrested with a gun are probably modest, since the probability that even a serious violent crime or property crime results in arrest is surprisingly low.

Giving out serious money for anonymous tips about illegal guns would increase the costs of carrying a gun and reduce the benefits; flashing a gun at a party might still score points, but it would now massively increase your legal risk.

These rewards might help undercut trust among gang members and could be particularly helpful in keeping guns out of schools. A bunch of logistical issues would need to be worked out, including how large the rewards would be (I think $1,000 or more wouldn't be crazy) and how police should respond to tips and confiscate guns while respecting civil liberties.

But this idea does have the big advantage of getting us out of the stale public debate about gun control, and it gives us a way to make progress on this major social problem right away.

Jesus "Manny" Castro Jr. became an active gang member at the age of twelve. After a brief incarceration, he joined Cornerstone Church of San Diego and now runs the GAME (Gang Awareness Through Mentoring and Education) program at the Turning the Hearts Center in Chula Vista, California.

Growing up in gangs and living the gang lifestyle, I have firsthand knowledge after seeing so many people die from gangs and guns! One great idea that can help to cut gun deaths in the U.S. is having the perpetrator's family be financially responsible for all emotional, mental, and physical damages that result from the victim's family's loss.

This should include (but not be limited to) garnishing their wages for their entire lives and having them pay all funeral arrangements and all outstanding debts. If the perpetrator is under eighteen, then not only will he have to do time in prison but his parents should also be required to serve at least half of the time on behalf of his crime. Everything starts and stops in the home!

The greatest way to make this happen is to make it law and set up organizations that educate parents on how to stop gun violence and clearly teach [their children] the consequences that result from gun violence. At Turning the Hearts Center, through our GAME program, we found that the young people we are working with care about their parents and what they think.

I get parents' input on what goes on at home so that I can implement and address their issues into our GAME curriculum. Kids have respect for their parents—and if parents knew that they would/ could do time for their children's behavior, perhaps they would stay more involved in their lives.

If the people in communities around the U.S. can model what we do at Turning the Hearts Center, we can make a difference in the world. Hard-core issues like gun deaths need hard-core consequences.

David Hemenway is a professor of health policy and director of the Harvard Injury Control Research Center at the Harvard School of Public Health, and author of Private Guns, Public Health.

Create the National Firearm Safety Administration.

A milestone in the history of motor vehicle safety in the United States, and the world, was the establishment (forty years ago) of what is now the National Highway Traffic Safety Administration (NHTSA). The NHTSA created a series of data systems on motor vehicle crashes and deaths and provided funding for data analysis. This enabled us to know which policies work to reduce traffic injuries and which don't.

The NHTSA mandated many safety standards for cars, including those leading to collapsible steering columns, seat belts, and airbags. It became an advocate for improving roads—helping to change the highway design philosophy from the "nut behind the wheel" to the "forgiving roadside." Improvements in motor vehicle safety were cited by the Centers for Disease Control and Prevention as a twentieth-century success story.

A similar national agency is needed to help reduce the public health problems due to firearms. Firearms deaths are currently the second leading cause of injury deaths in the United States; more than 270 U.S. civilians were shot per day in 2005, and 84 of those died. In response, Congress should create a national agency (as it did for motor vehicles) with a mission to reduce the harm caused by firearms.

The agency should create and maintain comprehensive and detailed national data systems for firearms injuries and deaths and provide funding for research. (Currently the National Violent Death Reporting System provides funding for only seventeen state data systems and no money for research.)

The agency should require safety and crime-fighting characteristics on all firearms manufactured and sold in the U.S. It should ban from regular civilian use products which are not needed for hunting or protection and which only endanger the public. It should have the power to ensure that there are background checks for all firearm transfers to help prevent guns from being sold to criminals and terrorists.

The agency needs the resources and the power (including standard setting, recall, and research capability) for making reasonable decisions about firearms. The power to determine the side-impact performance standards for automobiles resides with a regulatory agency, as does the power to decide whether to ban three-wheeled all-terrain vehicles (while allowing the safer four-wheeled vehicles).

Similarly, each specific rule regulating the manufacture and sale of firearms should go through a more scientific administrative process rather than the more political legislative process. It's time to take some of the politics out of firearm safety.

I ALMOST GOT SENT TO GUANTANAMO
(SDL)

I arrived at the West Palm Beach Airport yesterday, trying to make my way back to Chicago, only to see my flight time listed on the departure board as simply DELAYED. They weren't even pretending it was leaving in the foreseeable future.

With a little detective work, I found another flight that could get me home on a different airline. I bought a one-way ticket and headed for airport security.

Of course, the last-minute purchase of a one-way ticket sets off the lights and buzzers for the TSA. So I'm pulled out of the line and searched. First the full-body search. Then the luggage.

It didn't occur to me that my latest research was going to get me into trouble. I've been thinking a lot about terrorism lately. Among the things I had in my carry-on was a detailed description of the 9/11 terrorists' activities, replete with pictures of each of the terrorists and information about their background. Also, pages of my scribblings on terrorist incentives, potential targets, etc. It also was the first thing the screener pulled out of my bag. The previously cheery mood turned dark. Four TSA employees suddenly surrounded me. They didn't seem very impressed with my explanation. When the boss arrived, one of the screeners said, "He claims to be an economics professor who studies terrorism."

They proceed to take every last item out of both my bags. It has been a very long time since I cleaned out my book bag. This is a bag with twelve separate pockets, all of which are filled with junk.

"What is this?" the screener asks.

"It's a *Monsters, Inc.* lip gloss and key chain," I respond.

And so it went for thirty minutes. Other than the lip gloss, he was particularly interested in my passport (luckily it was really mine), my PowerPoint presentation, the random pills floating among the crevices of my bag (covered with lint and pencil lead from years in purgatory), and a beat-up book (*When Bad Things Happen to Good People*).

Finally satisfied that I was playing for the home team, he allowed me to board a plane to Chicago. Thank God I left at home my copy of the terrorist handbook that I recently blogged about, or I would have instead been flying straight to Cuba.

WEIRD BUT TRUE: *FREAKONOMICS*-FLAVORED COP SHOW BOUGHT BY NBC (SJD)

A few months back, Levitt and I were asked help put together a TV cop show based on the concepts of *Freakonomics*. The gist: a big-city police force, in crisis, hires a rogue academic to help get crime under control.

It struck us as a totally crazy but also strangely appealing idea. The concept had been hatched by Brian Taylor, a young exec at Kelsey Grammer's production company, Grammnet, which then partnered with Lionsgate; and the acclaimed writer Kevin Fox was brought on board. The show would be called *Pariah*.

A couple weeks ago, Levitt and I went to Los Angeles to help these guys pitch the show to the TV networks. Since we know nothing about TV, we tried to not talk too much and let Kevin, Brian, and Kelsey do their thing. And they did! Here's the news, from Deadline.com:

> NBC has bought *Pariah* . . . [T]he police procedural features characters inspired by the economic theory "Freakonomics" made popular by authors/economists Steven Levitt & Stephen Dubner. In *Pariah*, the Mayor of San Diego appoints a rogue academic with no law enforcement background to run a task force using Freakonomics-inspired alternative methods of policing.

Who knows how far this will go, but the ride has been fun so far. It was particularly enlightening to talk to Grammer about acting (he's currently starring in the high-end drama *Boss*, playing a Daley-ish mayor of Chicago). At one point, I asked him what it is about certain people that make their faces so appealing on the screen while other people, who might be better-looking or more attractive in some other way, just don't have that appeal.

He answered immediately: "Head size. Most successful actors have really big heads."

Physiologically, he meant. At least I think so.

Update: the collapse of this deal was fast even by Hollywood standards. After just a few conference calls, NBC informed the producers that they were changing direction, or that they had changed their minds, or they were changing their oil, or something. We are still waiting for our moment in the sun.

Chapter 10

More Sex Please, We're Economists

 *Of course we've blogged about sex but, weirdly, only other people's sex: not one of our eight thousand blog posts ever mentions our own sexual experiences. That said, we have had a few things to say about prostitution, STDs, and online dating.*

BREAKING NEWS: SOCCER FANS NOT AS HORNY AS PREVIOUSLY THOUGHT (SJD)

A few years ago, Germany legalized prostitution. It wasn't hard to surmise that this was meant to make Germany a bit more hospitable for World Cup fans. Brothels across the country staffed up and prepared for the World Cup boom—which, apparently, hasn't happened at all. It may well be that enough soccer fans already feel they're being screwed by the refs to bother going out at night and paying for it.

AN IMMODEST PROPOSAL: TIME FOR A SEX TAX?
(SJD)

Whereby:

- It has been observed that Democrats are generally in favor of taxation and Republicans are generally opposed to unnecessary sexual activity; and whereby:

- The unintended costs of sexual activity are unacceptably high, particularly in the political arena (c.f. Messrs. Clinton, Foley, Craig, and Edwards, to name just a fraction of the available examples); and whereby:

- The pursuit of sex is also extremely costly beyond the political realm, in terms of lost productivity, unwanted pregnancies, sexually transmitted diseases, and ruined marriages (and other committed relationships); and whereby:

- The federal government is now, as always, in need of more money;

It is hereby proposed that a new "sex tax" shall be levied upon the citizens of these United States.

Let it be clear that the aim of said tax is not to deter sexual activity itself, but rather to capture some of the costs imposed by certain extraneous sexual activity that, especially once made public, tends to divert precious resources from more worthy subjects; to this end:

- Married couples will receive a substantial credit for sanctioned, in-home sexual activity; and, conversely:

- The highest rates shall be paid for premarital, extramarital, and otherwise unusual or undesirable sexual activity; and:

- Sexual activity between members of the same gender; or activity between more than two participants; or in an airplane, on a beach, or in other "nontraditional" settings shall surely be taxed at a higher, though heretofore undetermined, rate.

Also to be determined is a scale for noncoital activity. The Internal Revenue Service shall be granted the full and complete authority to collect said tax. Furthermore:

- Payment of said tax, while voluntary, is no more voluntary than payments or credits on other tax-related activities such as: charitable contributions, business-related deductions, and cash received for goods and services, and is therefore expected to stimulate a very acceptable rate of compliance; additionally:

- Taxpayers will create a sexual paper trail that could prove advantageous in countless future scenarios, including but not limited to: employment, courtship, and participation in the political process; and:

- The typical IRS audit would become considerably more interesting for the auditor, and interesting work is a much-needed incentive to attract and retain qualified IRS employees.

It should be acknowledged that determining an acceptable name for said tax may be politically difficult, much like the "estate tax" and the "death tax" are in fact nomenclaturally diverse versions of the same tax used by opposing parties; candidates to consider include: the Family Creation Tax; the Extracurricular Intercourse and Lesser Sex Act Tax; and the Shtup Tax.

Furthermore:

- This is not the first time such a tax has been proposed in America; in 1971, a Democratic legislator from Providence, RI, named Bernard Gladstone proposed such a measure in his state; he called it "the one tax that would probably be overpaid," but sadly, the measure was promptly rejected as being in "bad taste," a position with which we summarily disagree; and whereby:

- A similar tax does have a historical (if fictional) precedence in the writings of one Jonathan Swift, who in *Gulliver's Travels* noted that in a place called Laputa, "The highest tax was upon men who are the greatest favourites of the other sex, and the assessments according to the number and natures of the favors they have received; for which they are allowed to be their own vouchers." And finally:

- It is unclear why both Swift and Gladstone proposed that the tax be levied solely upon males but, in light of recent and less-than-recent news events, they were probably 100 percent correct to have done so.

MORE SEX PLEASE, WE'RE ECONOMISTS (SJD)

Steven Landsburg is not known for having temperate opinions. An economics professor at the University of Rochester and a prolific writer, Landsburg regularly raises provocative theories: women choke under pressure, e.g., or miserliness is a form of generosity. He is the author of the books *The Armchair Economist* and *Fair Play*, which are in some ways direct forebears of *Freakonomics*. His latest is called *More Sex Is Safer Sex: The Unconventional Wisdom of Economics*. We asked him about the titular idea:

Q. Many of the stories in your book rest on the idea that people should alter their personal welfare for the greater good—for instance, STD-free men should become more sexually active to give healthy women disease-free partners. In our society, is it possible to put such ideas into practice?

A. Sure. We put such ideas into practice all the time. We think that the owners of polluting factories should give up some of their personal welfare (i.e., their profits) for the greater good, and we convince them to do that via tradable emissions permits (when we're being smart) or via clumsy regulations (when we're being dumb). We think that professional thieves should give up some aspects of their personal welfare (i.e., their thievery) for the greater good, and we convince them to do that with the prospect of prison terms.

Our personal welfare is almost always in conflict with the greater good. When something exciting happens at the ballpark, everyone stands up to see better, and therefore nobody succeeds. At parties, everyone speaks loudly to be heard over everyone else, and therefore everyone goes home with a sore throat. The one great exception is the interaction among buyers and sellers in a competitive marketplace, where—for fairly subtle reasons—the price system aligns private and public interests perfectly. That's a miraculous exception, but it is an exception. In most other areas, there's room to improve people's incentives.

One theme of *More Sex Is Safer Sex* is that some of those disconnects between private and public interests are surprising and counterintuitive. Casual sex is one of those examples. If you are a recklessly promiscuous person with a high probability of HIV infection, you pollute the partner pool every time you jump into it—and you should be discouraged, just as any polluter should be discouraged. But the flip side of that is that if you are a very cautious person with a low probability of infection—and a low propensity to pass on any infection that you do have—then you improve the quality of the partner pool every time you jump into it. That's the opposite of pollution, and it should be encouraged for exactly the same reasons that pollution should be discouraged.

I'M A HIGH-END CALL GIRL; ASK ME ANYTHING

In *SuperFreakonomics,* we profiled a high-end escort whose entrepreneurial skills and understanding of economics made her a financial success. We call her Allie, which is neither her real nor professional name. There was so much interest in Allie after the book came out that she agreed to field reader questions on the blog. They are paraphrased below, along with Allie's answers.

Q. Can you tell us how you became an escort, and what your family thinks—or knows—about your occupation?

A. My parents don't know about my work, or anything else about my sex life. I was a programmer when I decided to quit my job and become an escort. I was single and meeting people through a popular dating website. Finding someone "special" proved to be difficult, but I did meet many nice men. I had grown up in a repressive small town and I was, at that time, looking to understand my own sexuality. I have never attached my self-worth to some idea of virginity or monogamy, but I still had not really explored many of my desires. I was meeting people living alternative lifestyles, and, as I got to know them, the stereotypes that I had built up started to come apart. During this time I was in my mid-twenties, and I had an active sex life. One day I decided to enter the occupation of "escort" on an online instant messaging profile. Within seconds I had many responses, and after about a week of talking to a few people, I decided to meet a dentist at a hotel. The experience wasn't glamorous or nearly as sexy as I thought it might be.

However, I came away from the experience thinking, "It wasn't bad." I began to think that if I just had one appointment a month, I could pay my car loan with it, and have a little extra money. Eventually, I chose to work as an escort exclusively. At that time, the reason I gave up my programming job was the free time. I was caring for a family member with a serious illness—the free time and money was a huge benefit.

Q. Do you have any moral problem with what you do?

A. I do not have a moral problem with having sex for money, as long as it's safe, and between consenting adults. However, I have always been concerned about how the social and legal issues may affect my future and the people that I love.

Q. What kind of clients do you have?

A. My clients are generally white, married, and professional males, between forty and fifty years old, with incomes over $100,000 a year. They tend to be doctors, lawyers, and businessmen looking to get away for a few hours in the middle of the day.

Q. How many of your clients are married men?

A. Almost all of my clients are married. I would say easily over 90 percent. I'm not trying to justify this business, but these are men looking for companionship. They are generally not men that couldn't have an affair [if they wanted to], but men who want this tryst with no strings attached. They're men who want to keep their lives at home intact.

Q. What do your clients' wives know or think about them coming to you?

A. I rarely got the opportunity to find out if the wives were okay with it, but I did see several couples, so I assume *they* were okay with it.

Q. Do you know the real names of your clients?

A. Yes. Always. I insist that they give me their full names and their place of work so that I can contact them there before we meet. I also check their identification when we meet. I also use verification companies, which assist escorts in verification of clients. These companies do the verification of the client and put them in a database so that when the client wants to meet with a girl for the first time, he doesn't have to go through the verification process again. For a fee, I can call in and they will tell me if the client has a history of giving the girls problems, where he works, and his full name.

Q. What are your out-of-pocket costs?

A. $300 to $500 a month for my online basic ads
$100 a year for the website
$100 a month for a phone
$1,500 a year for photography
If I was touring then there were extra expenses such as travel costs, hotels, and more advertising costs.

Q. Do you have any regrets about your chosen profession?

A. Being an escort provided me with many opportunities that I'm not sure I would have gotten if I had not been an escort. That said, my choice to become an escort had a definite cost associated with it beyond the advertising, photos, and websites. I believe it is close to impossible to have a healthy relationship while working. So it can be a lonely life. In addition, hiding my job from my friends and family proved to be difficult for many reasons.

Q. How do you think prostitution would change if it were legalized? Would you want your own child to become a prostitute?

A. If the social and legal ramifications were gone, I think that being an escort might be like being a therapist (I have never been a therapist, so my knowledge is obviously limited). Like most escorts, a therapist sells his or her skills by the hour. A therapist also has to meet people for the first time not knowing who is walking in the door. Many have their own offices and work alone. In addition, the session is generally private and requires discretion. I imagine that many times therapists have patients that they like and some they don't. A therapist's revenue, like almost all other occupations, probably increases if the client feels that the therapist likes them. I don't mean to imply that I have the skills of a trained therapist, or to in any way demean what they do; I'm just observing some obvious similarities. If I had a child, I would hope that they would feel empowered, and have the opportunity to do whatever they desire to do, and that they would be in charge of their own sexuality. This job has its downsides, though, and can take a high toll on a person. I know that it's made many aspects of my life and my relationships more difficult. So, like any parent, I would always want more for my child than I had for myself.

Q. So are you in favor of legalization?

A. I feel that prostitution should be legal. If a couple meets for dinner and a bottle of wine, and have sex, that's a date. If they meet for dinner and a bottle of wine, and have sex, with money in an envelope left on the dresser, that's illegal. I realize that there are women in prostitution who are there because they feel like they have to be. These women work in a different part of the industry than I did. Many have drug or abuse issues, among other problems. I think, instead of spending time and finite resources on arresting and criminalizing these women, we should spend our resources on making sure that these women have other opportunities and a place to go for help. The women who don't want to be prostitutes shouldn't have to be, and they should be able to get the help they need. Women who want to be should be able to. I feel that no one should have to take a job to make a living that is against his or her own moral judgment.

Q. How would legalization affect your business model?

A. I'm sure it would cause me to lower my rates. I'm sure more people would take up prostitution as a profession, and I am sure more men would partake in the activity. That said, legalization does not remove all the barriers to entry. The job still would have a huge negative stigma associated with it, both for the escorts and the clients. In countries like Canada, enforcement of prostitution laws is extremely lax, and while rates are lower, they aren't wildly different. So there would still be men out there afraid of their wives finding out, and I still wouldn't want to share my job title with my family.

Q. Dubner and Levitt wrote that you have some economics training. Has that informed the way you think about your occupation?

A. Sure, here are some examples:
Dinner with friends = opportunity cost
Perfect information = review sites
Transaction cost = setting up an appointment
Repeated game = reputation
Product differentiation = not a blonde

Seriously, I wish I had known then what I know now.

FREAKONOMICS RADIO GETS RESULTS (SJD)

It's nice to have a podcast that is popular, but it's another thing to have a podcast that actually changes the world. Can you guess which of our recent episodes changed the world? Maybe the one about how drivers are legally allowed to kill pedestrians? The one called "Fighting Poverty with Actual Evidence"? Or maybe the one about how the avocados we buy in the U.S. help fund Mexican crime cartels?

Nope.

Here's an e-mail from a listener in Cincinnati named Mandi Grzelak:

> True story: while listening to your Feb. 6 podcast "What You Don't Know About Online Dating," I thought to myself, "I should try online dating!" After all, if NPR employees are on sites like OK-Cupid, I might have a shot with one! How amazing would that be?!
>
> Long story short: I signed up that afternoon, started with some e-mails, and went on my first

date (from the site, not ever) on Feb. 10. Tim and I have been inseparable ever since, bring each other endless amounts of happiness, and last night he proposed. I, obviously, said yes. We plan to elope in NYC this August, to avoid a large dramatic wedding. But you and your families are welcome to join us.

And it's all thanks to you!!!

We can die happy now. We may never move the needle on big social or policy issues, but as long as Mandi and Tim are together, we can take some satisfaction in that.

Chapter 11
Kaleidoscopia

The previous ten chapters have been organized according to theme, which makes this book of blog posts different from the blog itself—because it has no organization whatsoever. One of us decides to write something on a given day and then—click!—it's published. One post bears no relation to the others before or after it. This tends to give blog reading a kaleidoscopic quality—a quality we have tried to capture in this chapter, which has no discernible theme. A less charitable (or more discerning?) view might be that we found ourselves toward the end of this book with a supply of unrelated posts—a pile of miscellany—and decided to shoehorn them into a chapter that might have been more honorably titled "Miscellaneous." That would also be true.

SOMETHING TO THINK ABOUT WHILE YOU WAIT IN LINE AT KFC
(SDL)

I've loved the chicken at KFC ever since I was a kid. My parents were cheap, so KFC was splurging when I was growing up. About twice a year my pleading, perhaps in combination with a well-timed TV advertisement, would convince my parents to bring the family to KFC.

For as long as I have been eating KFC, the service has always been terrible.

Yesterday was a good example. I went with my daughter Amanda. From the moment we entered the store to the time we left with our food, twenty-six minutes had elapsed. The line was so slow inside the restaurant that we eventually gave up and went through the drive-thru. We eventually got our food, but no napkins, straws, or plastic ware. That was still better than the time I went to KFC only to be told that they were out of chicken.

What is so ironic about the poor service at KFC is that, at the corporate level, they seem to try so hard to achieve good service. The name tag on the guy behind the counter yesterday said that he was a "customer maniac," or something like that, as part of KFC's "customer mania." A few years back, I seem to remember they were focused on total quality improvement. At another point, I think they had posted on the wall a list of ten customer-oriented service mantras all workers were supposed to strive for.

So why is it that KFC's service remains so bad? I have two mutually consistent hypotheses as to why:

1. KFC doesn't have enough people working. The next time you are at McDonald's, count the number of workers. It always stuns me how many people are on duty. It is not uncommon to see fifteen to twenty people working at a time in a busy McDonald's. There seem to be many fewer people working at KFC. I think there were only four or five workers yesterday when I visited.

2. KFC's clientele is poorer than the customers at other fast-food outlets, and poor people are less willing to pay for good service. There is no question in my mind that service is generally terrible in places frequented by the poor. Whether it is because poor people care less about service, I'm not sure. I do know that I virtually never saw bad service in the entire year I spent visiting

Stanford, which I've always attributed to the fact that there are so many rich people in the area.

POSTMORTEM ON *THE DAILY SHOW* (SDL)

Well, I survived my appearance on *The Daily Show*. Some random reflections on the experience:

First, Jon Stewart sure seems like a fantastic guy. Smart, friendly, down-to-earth, funny the whole time on and off the camera. Maybe he should run for president sometime. I would vote for him. His only problem is that he is not so tall, and Americans grow their presidents tall.

Second, sitting in the studio, no matter how hard you to try, it is impossible to imagine that 2 million people are watching what you are doing (actually in my case 2,000,002 because my parents don't usually watch, but they were watching last night). Which is good if you are someone like me who is inherently anti-social and frightened by crowds. It certainly would be more nerve-racking to do an interview in front of a live audience of 2 million people stretched out over the Mall in Washington.

Third, television, except maybe *Charlie Rose*, is a terrible medium for trying to talk about books. I had a long interview—over six minutes—but Stewart was asking hard questions that I couldn't give real answers to (essentially he wanted me to explain regression analysis, but to do it in fifteen seconds). One key point in *Freakonomics* is that we try to show the reader how we get our answers, not just assert that we are right. On TV, there just isn't time to follow that path.

Fourth, it sure is nice to be in front of an audience that is dying to laugh at and respond to anything you say. (For instance, I'm not sure why, but the audience burst out laughing when I mentioned crack cocaine.) I wish the students in my 9 A.M. undergraduate lecture were so responsive. Of course, if my lectures were one-tenth as entertaining as *The Daily Show*, I bet my students would be plenty responsive.

DENTAL WISDOM
(SJD)

I really like my dentist, Dr. Reiss. He's in his late sixties, maybe even in his early seventies. To say that he knows his way around the mouth is an understatement. But that's not the only reason I like him. He recently told me how he solved a particular problem. Because he's getting on in years, a lot of his patients were asking him if he was retiring soon. He didn't like this question;

he's a guy who plays tennis twice a week, reads a million books, and keeps up on NYC's cultural and political scenes with great vigor. So instead of deflecting these annoying retirement questions one at a time, he found a relatively inexpensive way to signal his intentions to anyone who cared: he bought new furniture and equipment for his office. Suddenly the questions stopped.

As much as I generally dread the dentist's chair, I always wind up learning something. Yesterday was no exception. I was asking Dr. Reiss about the causes of tooth decay—genetics vs. diet, etc. etc.—when he began explaining why toothpaste is such a bogus product. Any claims that toothpaste makes about preventing decay, whitening teeth, etc., are totally fallacious, Dr. Reiss told me, because the FDA can't and won't allow the ingredients necessary to perform those chores in an over-the-counter product that children can easily get hold of. (That's why he recommends an antibacterial product like Gly-Oxide, a fairly foul-tasting but apparently effective means of killing the bacteria that cause decay.)

The other thing I learned yesterday was far more interesting, with far greater implications. He told me that tooth decay in general, even among wealthy patients, is getting worse and worse, particularly for people in middle age and above. The reason? An increased reliance on medications for heart disease, high cholesterol, depression, etc. Many of these medications, Dr. Reiss explained, produce dry mouth, which is caused by a constricted salivary flow; because saliva kills bacteria in the mouth, a lack of it means increased bacteria, which leads to increased tooth decay. Given the choice of taking these medicines versus having some tooth decay, I'm sure most people would still choose the medicines—but I am guessing that most people haven't thought about the link between the two.

Unfortunately, I have to go back to Dr. Reiss today. At least I'll probably learn a little something.

WHAT'S WITH ALL THE BULLSHIT?
(SDL)

Last year the book *On Bullshit* by philosophy professor Harry Frankfurt was a surprise bestseller, even reaching number one on the *New York Times* bestseller list for one week. That is an amazing commercial success for my friends at Princeton University Press.

The success of that book apparently inspired some other authors:

The golfer John Daly has an autobiography out this week entitled *My Life in and out of the Rough: The Truth About all the Bullshit You Think You Know About Me.* This book is published by HarperCollins, the same people who published *Freakonomics*. They were scared to death of the title "Freakonomics" when my sister Linda Jines first thought it up. I guess they have loosened up a bit.

Then there is *100 Bullshit Jobs . . . and How to Get Them* by Stanley Bing. This book was also released just this week. Guess who the publisher was? HarperCollins!

Then there was *The Dictionary of Bullshit*, published two weeks ago. At least that one wasn't HarperCollins. Be careful not to confuse *The Dictionary of Bullshit* with *The Dictionary of Corporate Bullshit*, published in February.

Then there is *Bullshit Artist: The 9/11 Leadership Myth*, which came out in paperback in March; *Bullets, Badges, and Bullshit*, also out in March; and *Another Bullshit Night in Suck City*, from last September.

Is this enough bullshit? Apparently not.

On the horizon for release later this month are *The Business of Bullshit* (not *The Dictionary of Business Bullshit*, although you could be forgiven for the mistake) and *Your Call Is Important to Us: The Truth About Bullshit*.

At least there is a few months' respite before *Hello, Lied the Agent: And Other Bullshit You Hear as a Hollywood TV Writer*, due out next September.

All I can say is, what the f---is going on here?

IF BARACK OBAMA IS AS GOOD A POLITICIAN AS HE IS A WRITER, HE WILL SOON BE PRESIDENT
(SDL)

This post was published on November 25, 2006, roughly five months before Obama announced he would run for president. It is one of the few correct predictions we have ever made.

This is not a political blog. I have no interest in politics. But I have been reading a great book that happens to be written by a politician.

The first time I heard of Barack Obama is when I saw his name springing up on those political signs people put in their front yards in election years. I knew nothing about him except that he was affiliated with the University of Chicago law school and he was running some hopeless campaign for the U.S. Senate. I figured the support he was getting in my hometown at the time was probably the only support he would get in the whole state. The city I lived in, Oak Park, is left-wing to the point of comedy at times. For instance, as you cross into the city, a sign informs you that you are entering a nuclear-free zone. I thought it would take little more than him having a name like "Barack Obama" to win over the folks in Oak Park.

I was not paying any attention to the Senate race when I happened to get called at random for a poll being conducted by the *Chicago Tribune*. They asked me who I was going to vote for in the upcoming Senate election. Just out of sympathy and loyalty to the University of Chicago, I said I would vote for Obama. That way, when the results of the poll came out, he would have a few percent of the electorate behind him and he wouldn't feel so bad. I was flabbergasted when I saw the results of the poll on the front page of the newspaper a few days later: Obama was in the lead for the Democratic primary! (This, of course, was well before he got tapped to give the keynote address at the Democratic National Convention.)

Because I am not very into politics, I didn't pay much attention to the Senate race (which eventually was a landslide with Obama crushing—of all people— Alan Keyes). I did see him give two speeches: the Democratic convention one and his acceptance speech the night he won. Both times, I felt like he cast some sort of spell over me. When he spoke, I wanted to believe him. I can't remember another politician ever having that effect on me. One friend, who knows Barack and who also knew Bobby Kennedy, said he had not seen anyone like Kennedy until he met Barack.

Anyway, all of this is just a long prelude to the fact that I picked up his book *The Audacity of Hope* and was blown away at how well written it is. His stories sometimes make me laugh out loud and at other times well up with tears. I find myself underlining the book repeatedly so I can find the best parts quickly again in the future. I am also almost certain he wrote the whole thing himself, based on people I know who know him. If you aren't giving *Freakonomics* as a Christmas gift this year, this would make a great gift.

I suppose I shouldn't be that surprised at what a good writer he is because I read his first book *Dreams from My Father* two years ago and loved that one as well. But unlike that first book, written fifteen to twenty years ago before he had political ambitions, I thought this new one would just be garbage. Rarely does a book so exceed my expectations. Also, I should stress that I don't agree with all his political views, but that in no way detracts from the enjoyment of reading the book.

If he has the same effect on others as he does on me, you are looking at a future president.

MEDICINE AND STATISTICS DON'T MIX
(SDL)

Some friends of mine recently were trying to get pregnant with the help of a fertility treatment. At great financial expense, not to mention pain and inconvenience, six eggs were removed and fertilized. These six embryos were then subjected to Preimplantation Genetic Diagnosis (PGD), a process that costs $5,000 all by itself.

The results that came back from the PGD were disastrous. Four of the embryos were determined to be completely non-viable. The other two embryos were missing critical genes/DNA sequences, which suggested that implantation would lead either to spontaneous abortion or to a baby with terrible birth defects.

The only silver lining on this terrible result is that the latter test had a false positive rate of 10 percent, meaning that there was a 1-in-10 chance that one of those two embryos might be viable.

So the lab ran the test again. Once again the results came back that the critical DNA sequences were missing. The lab told my friends that failing the test twice left only a 1-in-100 chance that each of the two embryos was viable.

My friends—either because they are optimists, fools, or perhaps know a lot more about statistics than the people running the tests—decided to go ahead and spend a whole lot more money to have these almost certainly worthless embryos implanted nonetheless.

Nine months later, I am happy to report that they have a beautiful, perfectly healthy set of twins.

The odds against this happening, according to the lab, were 10,000 to 1.

So what happened? Was it a miracle? I suspect not. Without knowing anything about the test, my guess is that the test results are positively correlated, certainly when doing the test twice on the same embryo, but probably across embryos from the same batch as well.

But the doctors interpreted the test outcomes as if they were uncorrelated, which led them to be far too pessimistic. The right odds might be as high as 1 in 10, or maybe something like 1 in 30. (Or maybe the whole test is just nonsense and the odds were 90 percent!)

Anyway, this is just the latest example of why I never trust statistics I get from people in the field of medicine, ever.

My favorite story concerns my son Nicholas:

Relatively early on in the pregnancy we had an ultrasound. The technician said that although it was very early, he thought he could predict whether it would be a boy or a girl, if we wanted to know. We said, "Yes, absolutely we want to know." He told us he thought it would be a boy, although he couldn't be certain.

"How sure are you?" I asked

"I'm about fifty-fifty," he replied.

IF YOU LIKE HOAXES . . .
(SJD)

. . . then you have to admit that this is a pretty good one: sending a piece of bogus research material to a biographer whom you happen to hate. In this case, the biographer is A.N. Wilson, who was writing a book about the poet John Betjeman. Wilson made use of the bogus letter, only to discover too late that the letter was fake—and that if you took the first letter of each sentence in the letter and added them one to the next, they would spell out this lovely message: "A.N. Wilson is a shit."

This reminds me of my first job in journalism, as an editorial assistant at *New York* magazine. Once or twice a week, it was my job to stay late to look over the closing page proofs to make sure there were no errors that the story editors, copy editors, or production editors had missed. The most important thing was to make sure that the "drop caps" (i.e., the jumbo capital letters that begin each new section of a magazine article) didn't inadvertently spell out something offensive. One night, while proofing an article about breast cancer, I found that the first four drop caps were *T, I, T,* and *S.* Yes, we changed them.

FROM GOOD TO GREAT . . .
TO BELOW AVERAGE
(SDL)

I almost never read business books anymore. I got my fill of them years ago when I was a management consultant before I went back and got a Ph.D.

Last week, however, I picked up *Good to Great,* by Jim Collins. This book is an absolute phenomenon in the publishing world. Since it came out in 2001, it has sold millions of copies. It still sells over three hundred thousand copies a year. It has been so successful that seven years later the book is still in hardcover. I've been hearing about it for years and never looked at it. People are always asking me about it. I figured it was about time I took a look.

The book focuses on eleven companies that were just okay, and then transformed themselves into greatness—where greatness is defined as a sustained period over which the stock dramatically outperformed the market and its competitors. Not only did these companies make the transition from good to great, but they also had the sort of characteristics which made them "built to last" (which is the title of an earlier book by Collins).

Ironically, I began reading the book on the very same day that one of the eleven "good to great" companies,

Fannie Mae, made headlines in the business pages. It looks like Fannie Mae is going to need to be bailed out by the federal government. If you had bought Fannie Mae stock around the time *Good to Great* was published, you would have lost over 80 percent of your initial investment.

Another one of the "good to great" companies is Circuit City. You would have lost your shirt investing in Circuit City as well, which is also down 80 percent or more.

Nine of the eleven companies remain more or less intact. Of these, Nucor is the only one that has dramatically outperformed the stock market since the book came out. Abbott Labs and Wells Fargo have done okay. Overall, a portfolio of the "good to great" companies looks like it would have underperformed the S&P 500.

I seem to remember that someone did an analysis of the companies highlighted in the classic Peters and Waterman book *In Search of Excellence* and found the same thing.

What does this all mean? In one sense, not much.

These business books are mostly backward-looking: What have companies done that made them successful? The future is always hard to predict, and understanding the past is valuable; on the other hand, the implicit message of these business books is that the principles that these companies use not only have made them good in the past, but position them for continued success.

To the extent that this doesn't actually turn out to be true, it calls into question the basic premise of these books, doesn't it?

This post was published in 2008. As of this writing, Fannie Mae is trading at just over $2 per share, down from nearly $80 in 2001, and Circuit City went bankrupt. The rest of the "Good to Great" companies are a mixed bag since 2008. Some have risen steeply (Kroger and Kimberly-Clark), others have fallen badly (Pitney Bowes and Nucor), while two of the eleven companies—Gillette and Walgreens—joined corporate forces (with, respectively, Procter & Gamble and Boots) and have had considerable success.

CUT GOD SOME SLACK
(SDL)

A while back, I blogged about how every third book had the word *bullshit* in its title. Happily, that trend faded. I could only find two books on Amazon released in the last year with *bullshit* in the title.

Now it seems that going after God is the hip thing to do. Daniel Dennett started the stampede with *Breaking the Spell.* Richard Dawkins followed with the best-seller *The God Delusion.* Then came *God, the Failed Hypothesis* by Victor Stenger and *God Is Not Great* by Christopher Hitchens.

Next up? *Irreligion* by John Allen Paulos (author of *Innumeracy*). I love the fact that the book's release date is December 26. What could be more fitting?

Here is what puzzles me: Who buys these books?

I'm not religious. I don't think much about God, except when I am in a pinch and need some special favors. I have no particular reason to think he'll deliver, but I sometimes take a shot anyway. Other than that, I'm just not that interested in God.

I'm definitely not interested enough to go out and buy books explaining to me why I shouldn't believe in God, even when they are written by people like Dennett and Dawkins, whom I greatly admire. If I were religious, I think it would be even more likely that I would go out of my way to avoid books telling me that my faith was misplaced.

So who is making these anti-God books bestsellers? Do the people who despise the notion of God have an insatiable demand for books that remind them of why? Are there that many people out there who haven't made up their mind on the subject and are open to persuasion?

Let me put the argument another way: I understand why books attacking liberals sell. It is because many

conservatives hate liberals. Books attacking conservatives sell for the same reason. But no one writes books
saying that bird-watching is a waste of time, because
people who aren't bird watchers probably agree, but
don't want to spend twenty dollars in order to read
about it. Since very few people (at least in my crowd)
actively dislike God, I'm surprised that anti-God books
are not received with the same yawn that anti-bird-
watcher books would be.

WHY I LIKE WRITING ABOUT ECONOMISTS
(SJD)

Over the years I have had the opportunity to write
about a great many interesting people. My mother had
an extraordinary (and long-buried) story to tell about
her religious faith. I've interviewed Ted Kaczynski, the
Unabomber; the rookie class of the NFL; a remarkable
cat burglar who stole only sterling silver.

But lately I have been writing about economists—
and, most fruitfully, with the economist Steve Levitt.
This is a whole new bag, and here's why.

A non-fiction writer like me, trained equally in
journalism and literature, is constrained by what his
subjects tell him. Yes, I am afforded great latitude in

surrounding a subject—if Ted Kaczynski won't discuss his trial, for instance, there are plenty of others who will—but I am gravely limited by what people will tell me and how they tell it.

The obvious fact is that when most people are being written about, they present themselves as well as they can. They tell the stories that make them look good, or noble, or selfless; some of the cleverer ones use self-deprecation to convey their excellence. Which leaves the writer in an unpleasant situation—dependent on anecdotes that may or may not be true, or complete, and which are told in order to paint a biased picture.

Here is where economists are different. Instead of using anecdotes to augment reality, they use data to assert the truth. That, at least, is the goal. Some of these truths can be uncomfortable. After I wrote about the economist Roland Fryer, he was assailed by fellow black scholars for having underplayed the degree to which racism afflicts black Americans. Steve Levitt's work with John Donohue on the link between *Roe v. Wade* and the drop in violent crime has made people of all political beliefs equally queasy.

But for me, the writer, this kind of thinking is a godsend—a way of looking at the world that's more long-sighted and unbiased than journalism typically affords.

Levitt likes to say that morality represents the way that people would like the world to work, whereas economics represents the way it actually does work. I don't have the mental horsepower to be the kind of economist that Levitt and Fryer are; but I feel lucky to have found a way to hitch my curiosities to their brains. In the parlance of economists, my skills and Levitt's are complementarities. Like most of the language of economics, the word itself is an ugly one; but, like a lot of economics itself, the concept is grand.

WHEN A DAUGHTER DIES
(MICHAEL LEVITT)

Steve Levitt wrote:

> My sister Linda passed away this summer. Nobody could love a daughter more than my father Michael loved Linda. My father, who is a doctor, was realistic from the start about what modern medicine might be able to do to save his precious daughter from cancer. Even with those low expectations, he was shocked at how impotent—and actually counterproductive—her interactions with the medical system turned out to be. Here, in his own words, is my father's poignant account of my sister's experience with medical care.

"Daddy, I am going to tell you something you are not going to want to hear. The MRI showed that I have two brain tumors." This verbal catastrophe is the telephone message that I (an elderly, practicing gastroenterologist) received from my previously healthy fifty-year-old daughter, who had just undergone a brain MRI for unsteady gait of one week's duration. A worrier and a pessimist, I feared the MRI might show multiple sclerosis. Metastatic brain tumors were outside even my fertile imagination. The date is August 9, 2012.

For unknown reasons, my daughter is transferred via ambulance to a local metropolitan hospital. In a one-hour period, the MRI result has converted my daughter into an ambulance case and me into a very nervous, distressed father. A total-body CT exam shows additional tumors in the neck, lungs, and adrenal glands with possible involvement of the liver. A relationship is formed with a local oncologist, the neck mass is biopsied, and my daughter is discharged to await biopsy results. Four days later, the biopsy is read as non-small-cell carcinoma of the lung. We are told that in young women who have never smoked, this tumor occasionally may have a favorable genotype that renders it susceptible to chemotherapy. An Internet check indicates that the favorable genotype is rare and "susceptible"—one of those relative terms employed in oncology.

A Greek aphorism warns, "Call no man happy until he is dead." A calamity that I hoped/presumed never would occur now seems likely—I am going to outlive one of my children. I am very unhappy, and my wife asks if we will ever be happy again.

My daughter needs local treatment of the brain tumors and systemic chemotherapy. She and her husband opt for care at a distant referral center. She promptly is seen by a neuro-oncologist at the referral center, and a PET scan confirms the widespread nature of the tumor. The following day gamma knife therapy is performed on the two major brain tumors, in the cerebellum and the frontal lobe. Nine days after the brain lesions were first observed, she leaves the referral center ostensibly in her usual good state of health (dexamethasone relieved the unsteady gait). Temporarily, I begin to eat and sleep once again. My daughter awaits a return visit to the referral center to discuss chemotherapy with a pulmonary oncologist. While I text or talk with her daily, I am completely unprepared for what I see when we meet five days later. She now looks ill. She is hoarse and short of breath with minimal exertion, and the neck mass seemingly has doubled in size. At this moment, the referral center notifies us that re-staining of the tumor indicates that it is of thyroid rather than lung origin. The appointment with the pulmonary oncologist is replaced by a visit with an endocrine-oncologist who recommends an adrenal biopsy to determine the differentiation of the metastatic tumor. Independent of the tissue of origin, it is apparent that a genetically altered monster is running rampant in my daughter's body.

No one is told that my daughter is ill other than her two siblings and my division director (to explain my absences) and an old friend who covers for me as a ward attending. This secrecy is attributable to my paranoia concerning public discussion of family health problems as well as the knowledge that my lacrimal glands are out of control. I know I will cry if anyone asks me about my daughter. An elderly doctor should not walk the halls of a hospital with tears streaming down his cheeks. In contrast, my wonderful, brilliant daughter is a model of self-control. No tears, no complaints. I suspect she has accepted the probable lethal outcome of her tumor and tolerates all the medical gyrations whirling about her to please her husband, son, and father. Is this the result of information from the Internet or have I non-verbally communicated my pessimism to her?

Six days after leaving the referral center in superficially good health, she returns in a wheelchair, short of breath at rest, and speaking in a whisper. Her oxygen saturation is 90 percent on room air. Since she has no stridor, the breathing problem apparently reflects tumor invasion of the lungs. Following the adrenal biopsy, her husband returns from the post-procedure observation room with the information that she has a rapid pulse. Until now I have remained the passive observer, but now am moved to intervene. I feel her pulse and her heart rate of about 145 is obviously irregular. I tell the nurse that I suspect atrial fibrillation and suggest that an EKG be obtained and the rapid intravenous infusion of saline be discontinued. To obtain an EKG, the Rapid Response Team must be called. This team arrives, an EKG shows atrial fibrillation, and her rate is slowed with beta and calcium channel blockers. Her blood oxygen saturation now is only 86 percent on five liters of oxygen. Her pulmonary function has deteriorated over eight hours. Can the monster tumor be expanding at this rate? To me, the rate-controlled atrial fibrillation is only a small problem on the rapid downhill progression of her malignant condition; to the young members of the Rapid Response Team, new-onset atrial fibrillation is the disease. I want to obtain a pulmonary arteriogram to rule out pulmonary emboli and sufficient oxygen to get her home, but both require transfer to the emergency room. I know this transfer is going to drag my exhausted daughter even deeper into the medical vortex of repeat histories, examinations, venesections, etc., but we acquiesce. A pulmonary arteriogram shows a massive tumor in the lung and no pulmonary emboli. The endocrine-oncologist visits her in the emergency room and patiently explains the need for determining the differentiation of the adrenal tumor to guide treatment. The response to my son-in-law's query if some treatment can be started immediately is that no treatment is better than misdirected treatment. She is scheduled to return to the referral center in four days to begin chemotherapy. I fear there will be no return visit.

Overnight admission to the hospital is recommended for "observation" and rest prior to the trip home. Fifty years of experience have taught me that admission to an academic hospital is not restful. I have stopped counting the patients who want to be discharged to get some rest. However, I fear she will not survive the trip home without supplemental oxygen, which only can be obtained via hospitalization.

She receives very little rest due to everything that happens on admission to a hospital—histories and physical exams by several residents, more blood tests, vital-sign checks seemingly every thirty minutes. I try to run interference—no echocardiogram, no anticoagulation, no cardiology consult, limit the vital-sign measurements, etc.—but by 8 A.M. she and her husband, who stayed in her room overnight, are exhausted.

My daughter and son want immediate discharge, but discharge requires an attending physician visit. I intercept the attending physician at about 10 A.M. and explain that my daughter has extensive metastatic carcinoma and all that is desired is rapid discharge with home oxygen. We are assured that this oxygen and discharge meds will be provided as rapidly as possible. Three hours later, we are still at the hospital. It is difficult to set up home oxygen on the weekend, and the pharmacy apparently has difficulty filling a prescription for a common drug. On my third visit to the hospital pharmacy, about 1.5 hours after they have received the prescription, I am informed it will be another thirty minutes until the medication will be ready. I insult the entire world of pharmacy with my query as to how hard can it be to put thirty tablets in a bottle.

At about 2 P.M., the oxygen and medications are ready. The only hurdle standing in the way of our departure is that my daughter fears she will be incontinent on the trip home. She needs an incontinence diaper. I then become an actor in a scene that must be played out many times each day in hospitals. At the nursing station, I explain the situation. The nurse says she will get the diaper, but she first makes a phone call that seems to go on forever (in reality, probably about three or four minutes). When she then begins to do some paperwork, I gently remind her that we need the diaper. She responds, "I have more patients to care for than just your daughter, Dr. Levitt." Of course she does, but I am only interested in the welfare of my daughter. We finally depart the hospital, no doubt with a well-deserved reputation for being a very difficult family.

Her condition continues to deteriorate at home, and it becomes apparent that she cannot tolerate a return trip to the referral center. Arrangements are made such that the local oncologist will administer the chemotherapy recommended by the endocrine-oncologist. My daughter can no longer speak and we exchange daily texts. On the day before she is to receive her first dose of chemotherapy (only eighteen days after the initial MRI), we exchange the following messages.

"When the chemotherapy does not work, you will have to finish the job."

"Be optimistic, I'll do whatever is necessary."

"Is that a yes?"

"Yes."

Exactly what I am going to do is not clear, but I intend to keep my promise.

The following morning, my son-in-law tells me she cannot get out of bed and coughs and gasps each time she tries to eat or drink. The "monster" now has destroyed her swallowing mechanism. It is apparent that she will not benefit from or tolerate chemotherapy. I talk with her local oncologist, who agrees to admission to the hospital via ambulance, presumably for comfort care. However, the ambulance driver has determined that her condition requires that she must be taken to the emergency room of the nearest hospital (less than ten minutes closer than the metropolitan hospital). I know she will not get comfort care at the local emergency room. I speak with the ambulance driver and forcefully tell him where I want my daughter/patient taken. The next thing I hear is that she is in the emergency room of the nearest hospital. When I arrive, she again has been through a series of tests, and yet another CT angiogram shows massive tumor invasion of the lung and no pulmonary emboli. She now is short of breath, on bi-pap and 100 percent oxygen. She is then transferred to the metropolitan hospital. Immediately upon arrival, my daughter asks for something which, with difficulty, I determine to be ice chips. I ask the nurse for ice chips. Her response is that nothing can be "administered" until ordered by the doctor. I tell her I am the doctor, and I want the patient to have ice chips. I am told I am not the admitting physician and cannot give orders. She ignores my request to show me the location of the ice machine.

Her oncologist arrives in a few minutes. Comparison of chest CTs shows that the undifferentiated tumor in her lung has doubled in size in less than three weeks. The hopelessness of the situation is discussed with her husband, and a decision is made with the assistance of a hospice physician to provide comfort care. She receives ice chips, and morphine is administered. About four hours later, she enters a peaceful coma and dies at 6:30 A.M. on August 29, just twenty days after the initial MRI demonstrated the brain tumors.

The purpose of this brief chronicle is not to criticize the practice of medicine. While I had several disagreements with non-physicians, the physicians who cared for my daughter, without exception, were very understanding and gave freely of their time. Each did everything possible to deal with her enormously aggressive malignancy. Rather, I have attempted to relate the experiences of a father/physician as he watches his daughter die of cancer. Her course was a testament to the limitations of medical care. In this era of molecular biology, the most valuable medication was morphine, a drug that has been available for almost two hundred years.

Although painful, I am capable of describing the events of my daughter's illness. When I try to describe my despair and grief, words fail.

LINDA LEVITT JINES, 1962–2012
(SDL)

It is with great sorrow that I share the news that my dear sister Linda Levitt Jines passed away last month after a short but valiant battle with cancer. She was fifty years old.

My very first instinct, as I sat down to try to eulogize Linda, was to call her to ask her to write it for me. Pretty much all my life, when faced with something that called for just the right words, that is what I've always done.

Most famously this happened when Dubner and I were halfway through writing a book that meandered from one topic to another and had no theme. Between the publisher, Dubner, and me, we had generated a list of perhaps fifteen terrible titles before we ran out of ideas. I knew with complete confidence that Linda would have the answer.

Indeed, it was just a matter of hours before she responded with a potential title: "Freakonomics." I liked it. Dubner wasn't sure. The publishers hated it. Our editor told us, "We gave you way too big of a book contract to call this thing *Freakonomics*!" In the end, though, *Freakonomics* won out, and it's a good thing it did. Without Linda's brilliant name, I doubt that anyone would have ever read the book. The name was a miracle.

Freakonomics wasn't the first time, or the last time, that I benefited from Linda's genius.

The first time I remember was when I was in seventh grade and she was in twelfth grade. I was the nerdiest, most socially awkward kid imaginable. She decided to take me on as a project. Then, as now, I was smart enough to know to listen to her. We became like best friends, and she gave me a complete makeover. She changed my clothes. She explained to me (nicely) how terrible and unlikable my personality was, and she helped me work on a new one. She introduced me to "cool" music—the first album I bought with my own money that year was U2's album *Boy*. After a few years of her tutelage, I was unrecognizable. I still couldn't get a date for another four or five years, but I was a lot more fun to be around. Looking through old scrapbooks, I stumbled upon an example of a note she had written to me that year, which pretty well captures the way her brain worked:

Dear Oink-baby,

The year is more than half over and it seems to me that you aren't yet hitched with a charming little sample of 7th grade womanhood. How do you resist their insidious allure? They're like the Sirens or the Lorelei! Doesn't your heart go all aflutter when you see those tempestuous maidens in repose (i.e. math class), natural patches of cochineal flitting across their cheeks as they contemplate various virtues of the opposite sex? Well, keep trying.

Your sister,
Linda

At my middle school, every student was required to memorize a short story or poem and recite it standing in front of the class. Two winners were chosen from each class, and they "got" to do their speech in front of a packed auditorium. I was a kid who almost never spoke. Nothing terrified me more than public speaking. I asked for Linda's advice. She told me she would take care of everything. She picked out a witty, lighthearted story for me. She practiced with me, coaching me on every line's delivery. But she knew it wouldn't be enough. The piece she had chosen was told by a girl. So she dug out one of her old dresses that would fit me. Then she grabbed one of my mom's blond wigs and put it on me. She taught me how to curtsy. She declared me ready. It says a lot about my faith in her that I dutifully dressed as a girl and delivered my speech just as she wanted it. I was selected to present in the packed auditorium. Against all odds, the shyest kid in the class, dressed in drag, took home the trophy. After that I never doubted her—I just did what she told me to.

When she wasn't busy pulling the puppet strings on my life, she was doing impressive things on her own behalf. After college she got a degree in journalism from Medill at Northwestern. She went into the creative side of advertising, landing a job at one of the top Chicago ad agencies. Amused by the goings-on at the shoot for the first commercial she wrote, she wrote a satirical piece about it for *Advertising Age*. She got fired the next day, which turned out to be a great career move. She got hired within days by a cross-town rival with a big salary increase.

Eventually she tired of advertising. In 1995, she called to tell me she was going to start an Internet business. Her idea: she was going to buy big jugs of fragrance oils used in soapmaking, pour the oil into little bottles with fancy labels, and resell them online. This sounded like the worst idea I had ever heard. First, no one was making money selling things over the Internet in 1995. Second, how could there possibly be enough soapmakers around to make good money? We all screamed at her not to waste her time. Seventeen years later, www.sweetcakes.com remains a thriving, highly profitable business. It never ceases to amaze me. Later, she started a second online business, www.yarnzilla. com. After *Freakonomics*, I started a little consulting company which eventually grew into the Greatest Good. Linda was the chief creative officer, her unique spirit imbuing everything we did.

And she did all this while she and her husband, Doug, raised the nicest, best-adjusted, most likable seventeen-year-old boy you could ever imagine, their son Riley. (Linda's expertise at turning boys into men clearly improved with practice over time, because even after her middle-school recrafting of me, I was nothing compared to Riley.) In addition to Doug and Riley, she leaves behind her parents Shirley and Michael, her sister Janet, and me, as well as many nieces and nephews who could never get enough of Auntie Lin.

Every time Linda entered the room, without even trying, she became the center of gravity. One of the people at the Greatest Good had never met Linda before. He walked into a conference room and all his co-workers were "grinning ear to ear." He wondered why. The answer was that Linda was there holding court.

The massiveness of her genius and creativity make the gaping hole of her absence all the greater.

Linda, we all miss you so much.

Chapter 12
When You're a Jet . . .

When you're a Freakonomist, you're a Freakonomist all the way. At least that is true for the two of us. We see economics everywhere we look, whether the subject is animated movies, baby formula, women's happiness, or pirates.

HOW MANY CHINESE WORKERS DOES IT TAKE TO SELL A CAN OF BABY FORMULA?
(SDL)

On a recent trip to China, I found that there were typically five people doing the job that one American would typically do. At our hotel, for instance, there was a floor monitor, whose main job, it seemed, was to press the button for the elevator. Maybe she also did other tasks I didn't notice, but she could always be relied on to hit that elevator button. In restaurants as well, servers were everywhere, seemingly one per table.

On the main street in Nanchang, there were perhaps two hundred people standing around with handwritten cardboard signs. I guessed maybe they were unemployed and looking for work. It turns out they actually were working, but I didn't realize it. Their job was to stand on the corner all day with a sign saying that they will buy used cell phones. Unfortunately for them, I saw perhaps three cell phones get sold in my week wandering up and down that street. It was the most competitive market I've ever seen. They must have been earning what they thought was a fair wage, though, or they wouldn't have been out there.

When I went to a large grocery store to buy a can of formula for my daughter Sophie, I thought I had observed the most extreme case of excess labor. As I searched the aisle for the exact type of formula she had been using in her orphanage, four young women very eagerly attempted to help me. At first I thought they were just shoppers trying to aid me. Eventually (they didn't speak English and I knew about fifty words of Mandarin) I realized they were working. Four of them huddled around me for roughly ten minutes before I finally purchased four dollars' worth of formula. It made absolutely no sense to me.

Only later, back at the hotel, did my Chinese guide explain what was going on. These women weren't employees of the grocery store; they were hired by rival formula companies to try to direct customers to their particular brand of formula! Which explains why they were all so cheerfully and persistently suggesting so many different kinds of formula to me. The store didn't care what kind of formula I bought—a sale was a sale. But to the formula manufacturers, stealing business from the rival brands was worth paying an employee to do.

WHY DO ANIMATED FILMS USE SUCH FAMOUS VOICES?
(SDL)

I took my four children to the movie *Coraline* this weekend. After the movie, I asked them how they liked it. Their four answers: "great," "good," "okay," and "Thank God it's over."

Coming from my kids, who always say the latest movie is their favorite, those are not very positive reviews.

I have never been in a movie theater full of kids as quiet as it was at *Coraline*. That quiet, along with the plodding pace of the movie, left plenty of time to ponder things.

First, I couldn't get over the fact that the name of one of the children in the movie was Wyborn, known as Wybie for short, as in "Why be born?" Wybie didn't seem to have any parents, although he did have a grandma who would yell for him from time to time. It made me think of the unwanted children/abortion argument in *Freakonomics*.

Second, two of the voices in this animated film were done by Dakota Fanning and Teri Hatcher. The last movie I saw was *Bolt,* with voices by Miley Cyrus and John Travolta. The list of stars who have recently lent their voices to animated movies goes on and on: Eddie Murphy, Dustin Hoffman, Cameron Diaz, John Goodman, etc.

Why do big-name stars so dominate the voices in animated films?

One hypothesis is that they are better than other people at doing the voices. I'm almost certain that is not correct. I have to believe that there are a group of voice actors and books-on-tape readers who don't have the faces to be movie stars but have great voices.

A second hypothesis is that the big stars don't charge much for their voices. According to what I've read in *The New York Times* and elsewhere, doing the voice for an animated film doesn't take much time or effort. If that is the case, then maybe the cost of the actors' voices is just a small part of the total cost of the movie; but I don't think that is the case, at least not always. I have read that Cameron Diaz and Mike Myers each got paid $10 million for their parts in *Shrek 2.*

A third explanation is that people really like to hear the voices of the stars. I tend to doubt that story as well. With a few notable exceptions, my guess is that audiences couldn't even identify the voices of the stars if they didn't see the credits.

A fourth hypothesis is one that sounds odd, but will be familiar to economists. Under this hypothesis, it isn't that famous actors are better at doing voices, or even that moviegoers like to hear their voices, or that stars are cheap. Rather, big-time actors are hired to read these parts precisely because they are expensive. In order to be willing and able to give multi-million-dollar deals to stars to do voices that a no-name could do for $50,000, a producer must be confident that the movie will be a big hit. Thus, the big star is hired solely to give a credible signal to outsiders that the producer thinks the movie will be a blockbuster.

Ultimately, I'm not sure any of these hypotheses really feel right to me.

WHY PAY $36.09 FOR RANCID CHICKEN?
(SJD)

An old friend came to town not long ago and we met for a late lunch on the Upper West Side. Trilby ordered a burger, no bread, with Brie; I ordered half a roasted chicken with mashed potatoes. The food was slow in coming but we had so much catching up to do that we didn't care.

My chicken, when it arrived, didn't look good but I took a bite. It was so rancid I had to spit it out into a napkin. Absolutely disgusting-gagging-rotten rancid. I summoned the waitress, a young and pretty redhead, who made a suitably horrified expression, then took the food away and brought back a menu.

The manager appeared. She was older than the waitress, with long dark hair and a French accent. She apologized, said the chefs were checking out the dish now, trying to determine if perhaps the herbs or the butter had caused the problem.

I don't think so, I told her. I think your chicken is rotten. I cook a lot of chicken, I said, and I know what rotten chicken smells like. Trilby agreed: you could smell this plate across the table, probably across the restaurant.

The manager was reluctant to concede. They had just gotten the shipment of chicken that morning, she said, which struck me as relevant as saying that "no, so-and-so couldn't have committed a murder today because he didn't commit one yesterday."

The manager left and, five minutes later, returned. You're right! she said. The chicken was bad. The chefs had checked the chicken, found it rotten, and were throwing it away. Victory! But for whom? The manager apologized again, asked if I'd like a free dessert or drink. Well, I said, first of all let me try to find some food on your menu that doesn't seem disgusting after that chicken. I ordered a carrot-ginger-orange soup, some french fries, and sautéed spinach.

Trilby and I then ate, fairly happily, though the taste of the rancid chicken remained with me; in fact, it remains with me still. Trilby had had a glass of wine before we ordered, and took another with her meal, sauvignon blanc. I drank water. When the waitress cleared our plates, she asked again if we wanted complimentary dessert. No, we said, just coffee.

As Trilby and I talked, I mentioned that I had not long ago interviewed Richard Thaler, the godfather of behavioral economics, which seeks to marry psychology and economics. Thaler and I had considered some small experiments at lunch—offering the waiter a gigantic tip, perhaps, in exchange for special considerations—but we didn't get around to it. Trilby was interested, so we kept talking about money. I mentioned the behavioralists' concept of "anchoring" (which used-car salesmen in particular know so well): establish a price that may be 100 percent more than what you need in order to ensure that you'll still walk away with, say, a 50 percent profit.

Talk turned to what we might say when our check came. There seemed two good options: "We don't care for any free dessert, thanks, but considering what happened with the chicken, we'd like you to comp our entire meal." That would establish an anchor at 0 percent of the check. Or this option: "We don't care for any free dessert, thanks, but considering what happened with the chicken, would you please ask the manager what you can do about the check." That would establish an anchor at 100 percent of the check.

Just then the waitress brought the check. It was for $31.09. Perhaps out of shyness, or haste, or—most likely—a desire to not appear cheap (when it comes to money, things are never simple), I blurted out Option 2: Please see what the manager "can do about the check." The waitress replied, smiling, that we had already been given the two glasses of wine for free. To me in particular this felt like slim recompense, since it was Trilby who had drunk the wine while it was I who still radiated with the flavor of rancid chicken. But the waitress, still smiling, duly took the check and headed toward the manager. She zipped right over, also smiling.

"Considering what happened with the chicken," I said, "I wonder what you can do about the check."

"We didn't charge you for the wines," she said, with great kindness, as if she were a surgeon who had thought she would have to remove both my kidneys but found instead that she had only had to remove one.

"Is that the best that you're prepared to offer me?" I said (still unable to establish an anchor at 0 percent).

She looked at me intently, still friendly. Here she was making a calculation, preparing to make the sort of gamble that is both financial and psychological, the sort of gamble that each of us makes every day. She was about to gamble that I was not the kind of person who would make a scene. After all, I had been friendly throughout our dilemma, never raising my voice or even uttering aloud the words *vomit* or *rancid*. And she plainly thought this behavior would continue. She was gambling that I wouldn't throw back my chair and holler, that I wouldn't stand outside the restaurant telling prospective customers that I'd gagged on my chicken, that the whole lot was rancid, that the chefs either must have smelled it and thought they could get away with it, or, if they hadn't smelled it, were so detached from their job that who knows what else—a spoon, a sliver of thumb, a dollop of disinfectant—might find its way into the next meal. And so, making this gamble, she said "yes": as in yes, that is the best that she was prepared to offer me. "All right," I said, and she walked away. I added a five-dollar tip to bring the total tab to $36.09—no sense penalizing the poor waitress, right?—walked outside, and put Trilby in a cab. The manager had gambled that I wouldn't cause trouble, and she was right.

Until now.

The restaurant, should you care to note, is called French Roast, and is on the northeast corner of Eighty-Fifth and Broadway, in Manhattan.

Last I checked, the roast chicken was still on the menu. Bon appétit.

PLEASE BUY GAS!
(SDL)

The e-mail reprinted below, which is circulating incredibly widely, may represent a new low in economic thinking. It declares September 1 "No Gas Day":

IT HAS BEEN CALCULATED THAT IF EVERYONE IN THE UNITED STATES AND CANADA DID NOT PURCHASE A DROP OF GASOLINE FOR ONE DAY AND ALL AT THE

SAME TIME, THE OIL COMPANIES WOULD CHOKE ON THEIR STOCKPILES.

AT THE SAME TIME IT WOULD HIT THE ENTIRE INDUSTRY WITH A NET LOSS OF OVER 4.6 BILLION DOLLARS WHICH AFFECTS THE BOTTOM LINES OF THE OIL COMPANIES.

THEREFORE SEPTEMBER 1ST HAS BEEN FORMALLY DECLARED "STICK IT TO THEM DAY" AND THE PEOPLE OF THESE TWO NATIONS SHOULD NOT BUY A SINGLE DROP OF GASOLINE THAT DAY.

THE ONLY WAY THIS CAN BE DONE IS IF YOU FORWARD THIS E-MAIL TO AS MANY PEOPLE AS YOU CAN AND AS QUICKLY AS YOU CAN TO GET THE WORD OUT.

WAITING ON THE GOVERNMENT TO STEP IN AND CONTROL THE PRICES IS NOT GOING TO HAPPEN. WHAT HAPPENED TO THE REDUCTION AND CONTROL IN PRICES THAT THE ARAB NATIONS PROMISED TWO WEEKS AGO?

REMEMBER ONE THING, NOT ONLY IS THE PRICE OF GASOLINE GOING UP BUT AT THE SAME TIME AIRLINES ARE FORCED TO RAISE THEIR PRICES, TRUCKING COMPANIES ARE FORCED TO RAISE THEIR PRICES WHICH EFFECTS [SIC] PRICES ON EVERYTHING THAT IS SHIPPED. THINGS LIKE FOOD, CLOTHING, BUILDING MATERIALS, MEDICAL SUPPLIES ETC. WHO PAYS IN THE END? WE DO!

WE CAN MAKE A DIFFERENCE. IF THEY DON'T GET THE MESSAGE AFTER ONE DAY, WE WILL DO IT AGAIN AND AGAIN.

SO DO YOUR PART AND SPREAD THE WORD. FORWARD THIS EMAIL TO EVERYONE YOU KNOW. MARK YOUR CALENDARS AND MAKE SEPTEMBER 1ST A DAY THAT THE CITIZENS OF THE UNITED STATES AND CANADA SAY "ENOUGH IS ENOUGH"

THANKS AND HAVE A WONDERFUL DAY :O}

Here is a (no doubt) partial list of totally idiotic mistakes in that e-mail:

1. If nobody buys gas today, but everybody drives the same amount, then it just means that we either had to buy more gas in anticipation of not buying any on September 1 or that we will buy more a few days later. So even if you believed this would take a $4.6 billion bite out of the oil companies that day, consumers would hand it right back over. If this was "No Starbucks coffee day" it might have some chance of mattering, because people buy and drink Starbucks coffee the same day, so a forgone cup of coffee today may never be consumed. But this is not true of gasoline, especially if no one is being asked to

reduce gas consumption. All you will get is longer lines at the pump the day after.

2. A one-day total boycott of gas would not reduce oil company bottom lines by anything like $4.6 billion, even if it was accompanied by a one-day moratorium on all gasoline use. Americans consume about 9 million barrels of gas a day. There are about 42 gallons in a barrel, so that equals 378 million gallons of gas sold a day in United States, or about one gallon per person. Toss in another 10 percent for Canada. At $3 a gallon, that is about $1.2 billion in revenues. Profit as a share of revenues in this industry is probably 5 percent or less, so the bottom-line impact is a max of $60 million—or about 1/100th of the stated number. And from point (1) above, even this is a gross exaggeration of the true impact.

3. One day of no purchasing of gasoline would certainly not cause the oil industry to choke on their stockpiles. Gasoline inventories in the U.S. are typically about two hundred million barrels, but right now they are on the low side—a big part of the reason why gas prices are high. Nine million extra barrels would create no problems whatsoever for stockpiles.

So everyone, please buy gas on September 1.

And if you ever have the bright idea to circulate an e-mail like this, at least tell people not to use gas, rather than not to buy gas.

This post was published in August, 2005, when the average price of regular gasoline in the U.S. was about $2.85 per gallon. As of this writing (January 2015), a gallon costs about $2.06, which gives people all the more reason to buy gas!

PIRATE ECONOMICS 101
(Q&A CONDUCTED BY RYAN HAGEN)

The crew of the *Maersk Alabama* recently survived an attack by pirates in Somalia and has returned home for a much-deserved rest. But with tensions growing between the U.S. and the ragtag confederation of Somali pirates, we thought it might be worth looking to the past for clues on how to tame the outlaw seas.

Peter Leeson is an economist at George Mason University and author of *The Invisible Hook: The Hidden Economics of Pirates.* Leeson agreed to answer some important piratical questions for us:

Q. *The Invisible Hook* is more than just a clever title. How is it different from Adam Smith's invisible hand?

A. In Adam Smith, the idea is that each individual pursuing his own self-interest is led, as if by an invisible hand, to promote the interest of society. The idea of the invisible hook is that pirates, though they're criminals, are still driven by their self-interest. So they were driven to build systems of government and social structures that allowed them to better pursue their criminal ends. They're connected, but the big difference is that, for Adam Smith, self-interest results in cooperation that generates wealth and makes other people better off. For pirates, self-interest results in cooperation that destroys wealth by allowing pirates to plunder more effectively.

Q. You write that pirates set up their own early versions of constitutional democracy, complete with separation of powers, decades before the American Revolution. Was that only possible because they were outlaws, operating entirely outside the control of any government?

A. That's right. The pirates of the eighteenth century set up quite a thoroughgoing system of democracy. The reason that the criminality is driving these structures is because they can't rely on the state to provide those structures for them. So pirates, more than anyone else, needed to figure out some system of law and order to make it possible for them to remain together long enough to be successful at stealing.

Q. So did these participatory, democratic systems give merchant sailors an incentive to join pirate crews, because it meant they were freer among pirates than on their own ships?

A. The sailors had more freedom and better pay as pirates than as merchantmen. But perhaps the most important thing was freedom from the arbitrariness of captains and the malicious abuses of power that merchant captains were known to inflict on their crews. In a pirate democracy, a crew could, and routinely did, depose their captain if he was abusing his power or was incompetent.

Q. You write that pirates weren't necessarily the bloodthirsty fiends we imagine them to have been. How does the invisible hook explain their behavior?

A. The basic idea is, once we recognize pirates as economic actors—businessmen really—it becomes clear as to why they wouldn't want to brutalize everyone they overtook. In order to encourage merchantmen to surrender, they needed to communicate the idea that, if you surrender to us, you'll be treated well. That's the incentive pirates give for sailors to surrender peacefully. If they wantonly abused their prisoners, as they're often portrayed as having done, that would have actually undermined the incentive of merchant crews to surrender, which would have caused pirates to incur greater costs. They would have had to battle it out more often, because the merchants would have expected to be tortured indiscriminately if they were captured.

So instead, what we often see in the historical record is pirates displaying quite remarkable feats of generosity. The other side of that, of course, is that if you resisted, they had to unleash, you know, a hellish fury on you. That's where most of the stories of pirate atrocities come from. This is not to say that no pirate ever indulged his sadistic impulses, but I speculate that the pirate population had no higher proportion of sadists than legitimate society did. And those sadists among the pirates tended to reserve their sadistic actions for times when it would profit them.

Q. So they never made anyone walk the plank?

A. There was no walking the plank. There's no historical foundation for that in seventeenth-or
eighteenth-century piracy.

Q. You write about piracy as a brand. It's quite a
successful one, having lasted for hundreds of
years after the pirates themselves were extermi-
nated. What was the key to that success?

A. There was a very particular type of reputation
that pirates wanted to cultivate. It was a delicate
line to walk. They didn't want to have a reputation for wanton brutality or complete madness.
They wanted to be perceived as hair-trigger
men, men on the edge, who if you pushed, if you
resisted, they would snap and do something horrible to you. That way, the captives they took
had an incentive to be very careful to comply
with all of the pirates' demands. At the same
time, they wanted a reputation for meting out
these brutal, horrible tortures to captives who
didn't comply with their demands. Stories about
those horrible tortures were relayed not only by
word of mouth, but by early-eighteenth-century
newspapers. When a former prisoner was
released, he would oftentimes go to the media

and provide an account of his capture. So when colonials read these accounts in the media, that helped institutionalize the idea of pirates as these men on the edge. That worked marvelously for pirates. It was a form of advertising performed by legitimate members of society that again helped pirates reduce their costs.

Q. What kinds of lessons can we draw from *The Invisible Hook* in dealing with modern pirates?

A. We have to recognize that pirates are rational economic actors and that piracy is an occupational choice. If we think of them as irrational, or as pursuing other ends, we're liable to come up with solutions to the pirate problem that are ineffective. Since we know that pirates respond to costs and benefits, we should think of solutions that alter those costs and benefits to shape the incentives for pirates and to deter them from going into a life of piracy.

THE VISIBLE HAND
(SDL)

Let's say you were in the market for an iPod and wanted to find a bargain, so you searched in a local online market like Craigslist to find one. Would it matter to you whether, in the photograph of the unopened iPod, the person holding the iPod (all you can see is their hand and wrist) was black or white? What if the hand holding the iPod had a visible tattoo?

I suspect that most people would say that the skin color of the iPod holder wouldn't matter to them. More people likely would say the tattoo might keep them from responding to the ad.

Economists have never liked to rely on what people say, however. We believe that actions speak louder than words. And actions certainly do speak loudly in some new research carried out by the economists Jennifer Doleac and Luke Stein. Over the course of a year, they placed hundreds of ads in local online markets, randomly altering whether the hand holding an iPod for sale was black, white, or white with a big tattoo. Here is what they found:

> Black sellers do worse than white sellers on a variety of market outcome measures: they receive 13% fewer responses and 17% fewer offers. These effects are strongest in the northeast, and are similar in magnitude to those associated with the display of a wrist tattoo. Conditional on receiving at least one offer, black sellers also receive 2–4% lower offers,

despite the self-selected—and presumably less biased—pool of buyers. In addition, buyers corresponding with black sellers exhibit lower trust: they are 17% less likely to include their name in e-mails, 44% less likely to accept delivery by mail, and 56% more likely to express concern about making a long-distance payment. We find evidence that black sellers suffer particularly poor outcomes in thin markets; it appears that discrimination may not "survive" in the presence of significant competition among buyers. Furthermore, black sellers do worst in the most racially isolated markets and markets with high property crime rates, suggesting a role for statistical discrimination in explaining the disparity.

So what can you conclude from this study? The clearest result is that if you want to sell something online, whether you are black or white, find someone white to put in the picture. I suppose you could say that advertisers figured this out long ago, and actually go one step further, making sure the white person is also a good-looking blond woman.

It is much harder, in this sort of setting, to figure out why buyers treat black and white sellers differently. As the authors note, there are two leading theories of discrimination: animus and statistical discrimination. By animus, economists mean that buyers don't want to buy from a black seller even if the outcome of the transaction will be identical. That is, buyers wouldn't like black sellers even if black sellers provided exactly the same quality as white sellers. With statistical discrimination, on the other hand, the black hand is serving as a proxy for some sort of negative: a higher likelihood of being ripped off, a product more likely to have been stolen, or maybe a seller who lives very far away so that it will be too much trouble to meet in person to do the deal.

The most impressive part of this paper by Doleac and Stein is their attempt to distinguish between these two competing explanations, animus versus statistical discrimination. How do they do it? One thing they do is to vary the quality of the advertisement. If the ad is really high quality, the authors conjecture, maybe that provides a signal that could trump the statistical discrimination motive for not buying from the black seller. It turns out that ad quality does not matter much for the racial outcomes, but possibly this is because the quality difference across the ads isn't great enough to matter. The authors also explore the impact of living in an area with more or less concentrated markets, and also across places with high and low property crime.

Black sellers do especially bad in high crime cities, which the authors interpret as evidence that it is statistical discrimination at work.

I really like this research a lot. It is an example of what economists call a "natural field experiment," which has the best of what lab experiments have to offer (true randomization) but with the realism that comes from observing people in actual markets, and with the research subjects unaware they are being analyzed.

BLACK AND WHITE TV
(SDL)

In *Freakonomics*, we mentioned in passing that blacks and whites in the U.S. have very different TV viewing habits. *Monday Night Football* is the only TV show that historically has been among the top ten in viewership for both blacks and whites. *Seinfeld*, one of the most popular white shows ever, was never in the top fifty for blacks.

So I was intrigued when I happened to see a recap of prime-time Nielsen ratings broken down by race.

The top ten shows for whites:

1. *CSI*
2. *Grey's Anatomy*

3. *Desperate Housewives*
4. *Dancing with the Stars*
5. *CSI: Miami*
6. *Sunday Night Football*
7. *Survivor*
8. *Criminal Minds*
9. *Ugly Betty*
10. *CSI: NY*

And for blacks:

1. *Grey's Anatomy*
2. *Dancing with the Stars*
3. *CSI: Miami*
4. *Ugly Betty*
5. *Sunday Night Football*
6. *Law and Order: SVU*
7. *CSI: NY*
8. *CSI*
9. *Next Top Model*
10. *Without a Trace*

If this one week of data is a good indicator (and I think it is), there has been a remarkable convergence in television viewing habits. A few years ago, almost all the top black shows featured predominately black characters and most were not even on the big-four networks. Now there is almost a perfect match between what blacks and whites are watching, and while many of these shows have black characters, none feature a predominantly black cast.

Does this convergence in TV viewing signal a broader pattern in cultural convergence? Probably not, but it is worth keeping an eye on.

Amid all the change, however, one thing seems to be as certain as death and taxes: both blacks and whites will watch football if you put in on in prime time.

HOW PURE IS YOUR ALTRUISM?
(SJD)

There have been a pair of huge natural disasters in recent weeks: a cyclone in Myanmar and an earthquake in China, each of which killed tens of thousands of people.

Have you written a check yet to donate to either cause? I seriously doubt it.

Why do I say that? Before looking at these recent tragedies, first consider three natural disasters from recent years, listed below with number of fatalities and amount of U.S. individual charitable donations (according to Giving USA):

1. Asian tsunami (December 2004)
   220,000 deaths
   $1.92 billion

2. Hurricane Katrina (August 2005)
   1,833 deaths
   $5.3 billion

3. Pakistan earthquake (October 2005)
   73,000 deaths
   $0.15 billion ($150 million)

Americans gave nearly three times as much money to Hurricane Katrina relief as they did for the Asian tsunami, even though the tsunami killed many, many more people. But this makes sense, right? Katrina was an American disaster.

Then along comes a terrible earthquake in Pakistan, killing seventy-three thousand people, and U.S. contributions are only $150 million, making the $1.92 billion given after the tsunami look very generous. That's only about $2,054 per fatality in Pakistan, versus an approximate $8,727 per fatality for the tsunami. Two faraway disasters both with huge loss of life—but with a huge disparity in U.S. giving. Why?

There are probably a lot of explanations, among them:

1. Disaster fatigue caused by Katrina and the tsunami; and

2. Lack of media coverage.

Do you remember coverage of the Asian tsunami? I am guessing you do, especially because in addition to hitting poor areas, it also struck high-profile resorts like Phuket. Do you remember coverage of Hurricane Katrina? Of course. But what about the Pakistan earthquake? Personally, I remember reading a couple of brief newspaper items but I didn't happen to see any coverage on TV.

Consider a recent paper by Philip H. Brown and Jessica H. Minty called "Media Coverage and Charitable Giving After the 2004 Tsunami." Here is their rather startling—if sensible—conclusion:

Using Internet donations after the 2004 tsunami as a case study, we show that media coverage of disasters has a dramatic impact on donations to relief agencies, with an additional minute of nightly news coverage increasing donations by 0.036 standard deviations from the mean, or 13.2 percent of the average daily donation for the typical relief agency. Similarly, an additional 700-word story in *The New*

York Times or *Wall Street Journal* raises donations by 18.2 percent of the daily average. These results are robust to controls for the timing of news coverage and tax considerations.

And what causes one disaster to get a lot of coverage while another doesn't? Again, there are probably a lot of factors, foremost among them the nature of the disaster (i.e., how dramatic/telegenic is it?) and location. Getting back to the recent disasters in Myanmar and China, I'd say there are a few other things worth considering:

1. We are in a season of heavy political coverage in the U.S., which is hard to dislodge from the airwaves.

2. Covering faraway disasters is time-consuming and expensive, which becomes doubly prohibitive when media outlets are in cost-cutting mode.

3. Neither Myanmar nor China (nor Pakistan) has what one would consider a very high Q Score among Americans. I am guessing that most Americans couldn't find Myanmar on a map, and if they have any impressions about the country at all, they are not good impressions (think "military junta").

Indeed, donations to Myanmar so far are very, very low. Considering how unevenly disaster aid is often distributed, maybe this isn't so terrible. But still: if you are the kind of person who donates money to people in need, isn't the family of a cyclone victim in Myanmar as worthy of your charity as anyone else? The political or narrative forces of a disaster shouldn't change our response to the need, should they?

We might like to think that we donate almost blindly, depending on need rather than our own response to the particulars of a disaster. But the growing economics literature on charitable donations shows that isn't the case. In a narrow but very compelling piece of research, John List argued that if you are trying to solicit donations door to door, the single best thing you can do to get large donations is to be an attractive blond woman.

I thought of this research when the NFL was raising money in a weekend telethon after Hurricane Katrina. Between games and during half times, the league had star players manning the phones. Relative to how many people watch football, the amount of money the league raised was pitifully small. I wondered if they would have done a lot better by using cheerleaders to solicit donations instead of the players.

So given the particulars of the disasters in Myanmar and China, as tragic as they are, I feel pretty confident in predicting that U.S. charitable contributions in each case won't be very large. It may be that the only kind of altruism that truly exists is what economists like to call "impure altruism." Does this mean that human beings are shallow and selfish—that they only give to a cause when it is attractive to them on some level? Will the future produce some sort of "disaster marketing" movement in which aid agencies learn to appeal to potential contributors?

THE ECONOMICS OF STREET CHARITY
(SJD)

Not long ago I was having dinner with Roland Fryer and our significant others. For some reason, talk turned to street charity. The conversation was so interesting that I thought we'd pose a street charity question to a few other folks. Their answers are presented below (and, FWIW, at the very end you can see what Roland and I thought).

The participants are: Arthur Brooks, who teaches business and government at Syracuse and is the author of *Who Really Cares: The Surprising Truth About Compassionate Conservatism*; Tyler Cowen, an economist at George Mason who writes books and maintains the Marginal Revolution blog; Mark Cuban, the multifaceted entrepreneur and Dallas Mavericks owner; Barbara Ehrenreich, author of the low-rent classic *Nickel and Dimed* and many other works; and Nassim Nicholas Taleb, the noted flâneur and author of *The Black Swan* and *Fooled by Randomness.*

Here is the question we put to each of them:

You are walking down the street in New York City with ten dollars of disposable income in your pocket. You come to a corner with a hot-dog vendor on one side and a beggar on the other. The beggar looks like he's been drinking; the hot-dog vendor looks like an upstanding citizen. How, if at all, do you distribute the ten dollars in your pocket, and why?

Arthur Brooks

We face this situation all the time—both literally and figuratively. If you live in a city, you are frequently confronted by needy winos. Do you give to them, or not? In your heart, you fear that they will just ruin their lives further with your pocket change. But it feels hardhearted not to give.

This dilemma goes beyond just how we treat the homeless. In our public policies, we see parts of the population which, we fear, might become dependent on the government "dole" if we provide that kind of help to people in need. Some even argue that whole nations can lose their self-sufficiency through foreign aid. This is why we have metaphors about giving fish versus teaching people to fish, and so forth.

Furthermore, some people worry a lot about the dignity of folks in need. For some, that means we should give them whatever they ask for. For others, it means charity is degrading and no good, and should be replaced totally by government programs.

As the Inuits say, "Gifts make slaves, as whips make dogs."

So how does all this help me figure out what to do as I approach the tipsy beggar and the upstanding hot-dog vendor? I have to figure out whether I care about a) the desires and sovereignty of the beggar; and b) the impact and effectiveness of my gift to do good in the world. There are four possibilities, with four different associated actions:

1. I care about the beggar's sovereignty, but not the impact of my gift. I give him some cash, which he will probably spend on booze. But hey, we all have free will, right? I didn't force him to buy booze instead of food.

2. I care about the impact of my gift but not the beggar's sovereignty. I buy him a hot dog—or better yet, I donate the money to a cause to help the homeless.

3. I care about both the beggar's sovereignty and the impact of my gift. This is the toughest case, and usually involves the futile exercise of trying to convince the beggar to "get some help." Imagine trying to have an intervention on the street.

4. I don't care about either the beggar's sovereignty or the impact of a gift. This is the easiest case of all. I buy myself a hot dog and ignore the wino. Put some kraut on that and give me a Diet Pepsi, too.

Which is my choice? I usually take number two, unless I'm feeling really lazy or I'm with somebody who knows I write books about charity—in which case I sometimes choose number one.

Tyler Cowen

I'm not keen on giving the money to the beggar. In the long run, this only encourages more begging. If you imagine a beggar earning, say, $5,000 a year, over time would-be beggars would invest about

$5,000 worth of time and energy into being beggars. The net gain is small, if indeed there is one. It is rumored that in Calcutta people cut off body parts to be more effective beggars; that is a polar example of this phenomenon. I explain this logic in more detail in my book *Discover Your Inner Economist.*

Oddly, the case for giving to the beggar may be stronger if he is an alcoholic. Alcoholism increases the chance that he is asking for the money randomly, rather than pursuing some well-calculated strategy of wastefully investing resources into begging. But in that case, I expect the gift will be squandered on booze, so I still don't want to give him the money.

If I liked hot dogs, I would buy a hot dog from the vendor rather than giving him the money for free. At the end of the day, he'll probably throw out food. He's going to get the money in any case, so why waste a hot dog?

A third option, only implied in the question, is to simply rip up the money. This will make the currency of others worth proportionately more and spread the gains very broadly. Since many dollar bills are held by poor foreigners (most of all in Latin America), the gains would go to those who are able to save in terms of dollars. This would include many hardworking poor people, a group I regard as worthy recipients.

I have two worries about this option, however. First, drug dealers and other criminals hold lots of cash—why should I help them out? Second, the Federal Reserve might (if only in the probabilistic sense) reverse the effect of my actions by printing more currency.

The bottom line: buy a hot dog.

The second bottom line: don't exercise your charity in New York City.

Mark Cuban

I keep the money in my pocket and keep walking because I have no reason to just hand over money on a street corner.

Barbara Ehrenreich

Could we first dispense with the smarmy connect-the-dots answer this question seems to cry out for? That is, that I'd use the ten dollars to buy a hot dog for the beggar and perhaps give the change to the vendor as a tip, thus rewarding a hardworking citizen while assuring that the shiftless beggar does not get the wherewithal for another drink—while of course giving me a nice little hit of middle-class self-righteousness.

Although I'm atheist, I defer to Jesus on beggar-related matters. He said, if a man asks for your coat, give him your cloak, too. (Actually, he said if a man "sue thee at the law" for the coat, but most beggars skip the legal process.) Jesus did not say: First, administer a Breathalyzer test to the supplicant, or, first, sit him down for a pep talk on "focus" and "goal-setting." He said: Give him the damn coat.

As a matter of religious observance, if a beggar importunes me directly, I must fork over some money. How do I know whether he's been drinking or suffers from a neurological disorder anyway? Unless I'm his parole officer, what do I care? And before anyone virtuously offers him a hot dog, they should reflect on the possibility that the beggar is a vegetarian or only eats kosher or halal meat.

So if the beggar approaches me and puts out his hand, and if I only have a ten-dollar bill, I have to give it to him. It's none of my business whether he plans to spend it on infant formula for his starving baby or a pint of Thunderbird.

Nassim Nicholas Taleb

This question is invalid and answers to it would not provide useful information. Let me explain:

When I recently had drinks and cheese with Stephen Dubner (I ate 100 percent of the cheese), he asked me why economics bothers me so much as a discipline, to the point of causing allergic reactions when I encounter some academic economists. Indeed, my allergy can be physical: recently, on a British Airways flight between London and Zurich, I found myself seated across the aisle from an Ivy League international economist dressed in a blue blazer and reading the *Financial Times.* I asked to be moved and preferred a downgrade, just to breathe the unpolluted air of economy class. My destination was a retreat in the Swiss mountains, in a setting similar to that of Mann's *Magic Mountain,* and I wanted nothing to offend my sensibility.

I told Stephen that my allergy to economists was on moral, ethical, religious, and aesthetics grounds. But here is another, central reason: what I call "ludicity," or the "ludic fallacy" (from the Latin *ludes,* meaning "games"). It corresponds to the setup of situations in academic-style multiple-choice questions, made to resemble "games" with crisp, unambiguous rules. These rules are divorced from both their environment and their ecology. Yet decision making on Planet Earth does not usually involve exam-style multiple-choice questions isolated from their context—which is why school-smart kids don't do as well as their streetwise cousins. And, if people often sometimes appear inconsistent, as shown in many "puzzles," it is often because it is the exam itself that is wrong. Dan Goldstein calls this problem "ecological invalidity."

So ecologically, in real life, we act in a different way depending on the context. As such, if I were to ecologize this question, I would answer it as follows: should I be walking down the street in New York City, I would rarely be faced with a mission to disburse ten dollars—I would usually be thinking about my next book, or how to live in an economist-free society (or one free of analytical philosophers). And my reaction would depend on the sequence: beggar first or street vendor first.

Should I run into a beggar, I would try to resist giving him money (I give enough to anonymous people via charity), but I am certain that I may not succeed. I need to be actually facing a drunk beggar for that. My reaction also depends on whether I would have been exposed to images of starving children before the encounter—these would sensitize me. And do not underestimate personal chemistry. I may give him a lot more than ten dollars if he reminds me of my beloved great-uncle, or cross the street if he remotely looks like the economist

Robert C. Merton. Of course, should you debrief me after the fact, I would never give "chemistry" as a reason for my choice, but rather some theoretical smart-sounding narrative.

Now, my airplane story has a twist. On the same British Airways route between Switzerland and London, I once sat next to another economist who was perhaps the first to uncover such ecological invalidity. His name was Amartya Sen, and he introduced himself as a philosopher, not as an economist. Furthermore, he looked physically similar to the first economist (though he did not wear a blue blazer). I was proud to breathe the same air as Sen.

For the record, when Fryer and I had our conversation on this subject a few weeks ago, neither of us put as much effort into it as the folks above (with the exception, perhaps, of Cuban).

My position was that panhandling is almost universally inefficient and a nuisance to boot; and, as I prefer to reward good behavior rather than punish bad, I would give the hot-dog man some or all of my money. He's the one, after all, who's out there every day providing a service, having to pay taxes, licensing fees, etc. The panhandler, meanwhile, has far more efficient and effective options for getting food and shelter than getting a random few dollars from the likes of me, and the more I give, the more I ask him to spend his time on the street.

Roland, meanwhile, said he'd give his ten dollars to the panhandler: it was such a small amount, he said, that it would make more of a marginal impact on the panhandler than the hot-dog guy.

BRIBING KIDS TO TRY HARDER ON TESTS
(SDL)

We use direct financial incentives to motivate so many different activities in life. No one expects workers in a fast-food restaurant to flip burgers for free. No one expects teachers to show up and teach without getting paid. But when it comes to kids in school, we think that the distant financial rewards they will earn years or decades later should be enough to motivate them, even though for most kids a month or two feels like an eternity.

To learn a little more about whether kids' school effort responds to financial incentives, I carried out a series of field experiments with John List, Susanne Neckermann, and Sally Sadoff. We recently wrote up the results in a working paper.

Unlike most previous studies involving kids, schools, and payments, in this research we aren't trying to get kids to study hard or learn more. We were going after something even more simple: just get the student to try hard on the test itself. So we don't tell the kids about the financial reward ahead of time—we just surprise them right before they sit down to take the test by offering them up to twenty dollars for improvements.

To see any gains from the financial incentives, the students need to know that they will be paid right away. If instead we tell them we will pay them one month later, they don't do any better than with no incentives at all. This is bad news for those who argue that payoffs that come years or decades in the future are sufficient to motivate students.

The very best results come when we give the students the money before the test, and then we take the money back if they don't meet the standards. This result is consistent with what psychologists call "loss aversion."

With young kids, it is a lot cheaper to bribe them with trinkets like trophies and whoopee cushions, but cash is the only thing that works for the older students.

It is remarkable how offended people get when you pay students for doing well—so many negative e-mails and comments. Roland Fryer endured the same onslaught as he has experimented with financial incentives in cities around the U.S.

Perhaps the critics are right and the reason I'm so messed up is that my parents paid me twenty-five dollars for every A that I got in junior high and high school. One thing is certain: since my only sources of income were those grade-related bribes and the money I could win off my friends playing poker, I tried a lot harder in high school than I would have without the cash incentives. Many middle-class families pay kids for grades, so why is it so controversial for other people to pay them?

THE SALMON IS DELICIOUS: AN EXAMPLE OF INCENTIVES AT WORK
(SDL)

A group of us went out for dinner the other night at a reasonably fancy restaurant. As we looked over the menu, the waitress was kind enough to let us know that the salmon was particularly delicious. We might also want to try the artichoke dip, she told us. It was her personal favorite.

Alas, our preferences were not easily swayed. None of us ordered the salmon, and there was little interest in the artichoke dip. As the waitress collected the menus, she inquired once again as to whether we might not want to give the artichoke dip a chance. Half joking, one of us asked her if there was a particular reason why she wanted us to try it.

No doubt sensing that she was talking to a bunch of nerdy economists who would appreciate the truth, she answered honestly: the chef had created a new dessert (and she loves dessert). Whichever member of the waitstaff sold the greatest number of artichoke dips and salmon entrées that night would collect a healthy portion of the new dessert for free. We duly rewarded the restaurant's creative approach to incentives by adding an artichoke dip to our order.

Later in the meal, I asked her whether the restaurant incentivized the waitstaff to sell particular products frequently. She mentioned that on an earlier occasion, they had offered a $100 prize to the person who sold the most orders of a certain dish.

"Wow," I said. "That hundred dollars must have really lit a fire under you."

"Actually," she replied, "I'm more excited about the dessert."

Chalk up another victory for non-pecuniary incentives.

SHRIMPONOMICS

(SDL)

I recently posed a simple question on the blog: "Why are we eating so much shrimp?" (Between 1980 and 2005, the amount of shrimp consumed per person in the U.S. has nearly tripled.) I didn't expect more than one thousand responses!

I asked the question because Shane Frederick, a marketing professor at MIT's Sloan School, had contacted me with an intriguing hypothesis. He wrote about a striking regularity in the responses he got when he asked different people why we are eating so much shrimp:

Psychologists (indeed, probably all non-economists) give explanations that focus on changes in the position of the demand curve—changes in preferences or information, etc., like:

1. People are becoming more health-conscious and shrimp is healthier than red meat;

2. Red Lobster switched ad agencies, and their ads are now working;

And so on. Economists, by contrast, tend to give explanations that focus on "supply," like:

1. People have designed better nets for catching shrimp;

2. Weather conditions in the Gulf have been favorable for shrimp eggs;

And so on.

I found Shane's hypothesis compelling. When I teach intermediate microeconomics, the students seem to understand demand a lot more easily than supply. Most of us have a lot more experience being consumers than producers, so we tend to view things through the lens of demand rather than supply. We need to have an appreciation of supply factors trained into us by economists.

My colleagues generated some confirmatory evidence regarding Shane's hypothesis. All eight of the University of Chicago economists to whom I posed the shrimp question thought the answer had something to do with producing shrimp more efficiently—i.e., supply-based explanations.

Which led me to open up the question to blog readers to see what their responses would look like. With the help of Pam Freed (a Harvard undergrad who plans to be an economics major and first gave a "demand" explanation, but quickly switched to a "supply" story in response to my withering stare), we cataloged the first five hundred blog comments we received.

Well, Shane, I am sorry to report that your hypothesis only did so-so in the data.

There were 393 usable observations (107 of you didn't follow the directions).

First, the good news for the hypothesis. As Shane conjectured, non-economists (i.e., anyone who didn't major in economics) mostly thought that we are eating more shrimp because of demand-based reasons (e.g. the movie *Forrest Gump,* a rise in the number of vegetarians who will eat shrimp, etc.). Fifty-seven percent of non-econ majors gave only demand stories, versus 24 percent who gave only supply stories. The rest had a mix of supply and demand explanations.

Where the theory didn't do so well, however, is that the 20 percent of the respondents who were economics majors didn't look all that different from everyone else. Roughly 47 percent of the econ majors exclusively gave demand stories, and 27 percent only supply. (Economics majors were more likely to give both supply and demand stories.)

In fairness to Shane, there is a big difference between being an economics professor and having an undergraduate major in economics. Indeed, the similarity between economics majors and everyone else is, perhaps, an indication that our current curriculum for teaching economics doesn't do a great job of instilling students with good economic intuition—or at least whatever economic intuition my colleagues have.

Who thinks least like the academic economists? That prize goes (no surprise) to English majors and (more of a surprise) engineering majors, who together combined to give forty-nine responses that overwhelmingly touted demand explanations.

Interestingly, women in general were only half as likely to give supply explanations as were men. I will leave you to ponder the causes and implications of that result.

So why did shrimp consumption rise so much?

I'm not exactly sure, but one key factor is that prices have dropped sharply. According to one academic article, the real price of shrimp fell by about 50 percent between 1980 and 2002. When quantity rises and prices are falling, that has to mean that producers have figured out cheaper and better ways to produce shrimp. An article in *Slate* argues there has been a revolution in shrimp farming. Demand factors may also be at work, but they don't seem to be at the heart of the story.

So for the diligent few who have actually read all the way to the end of this post, here is another question: In stark contrast to shrimp consumption, the amount of canned tuna eaten has been steadily falling; is that due to changes in supply or demand?

WHY ARE WOMEN SO UNHAPPY?
(SDL)

I saw Justin Wolfers a few weeks back, and I joked with him that it had been months since I'd seen his research in the headlines. It didn't take him long to fix that—he and Betsey Stevenson, his partner in life and economics, made the news twice last week. The first was in the form of a *Times* op-ed pointing out that the media has totally misinterpreted newly released statistics on divorce. While the reports trumpeted the new data as evidence that Americans today are more likely than ever to get divorced, Stevenson and Wolfers show that this pattern is purely an artifact of a change in how the data are collected. In fact, fewer people today are getting married, but the ones who do are more likely to stay together.

In addition, Stevenson and Wolfers released a new study, "The Paradox of Declining Female Happiness," that is bound to generate a great deal of controversy. By almost any economic or social indicator, the last thirty-five years have been great for women. Birth control has given them the ability to control reproduction.

They are obtaining far more education and making inroads in many professions that were traditionally male-dominated. The gender wage gap has declined substantially. Women are living longer then ever. Studies even suggest that men are starting to take on more housework and child-raising responsibilities.

Given all these changes, the evidence presented by Stevenson and Wolfers is striking: women report being less happy today than they were thirty-five years ago, especially relative to the corresponding happiness rates for men. This is true of working women and stay-at-home moms, married and single women, the highly educated and the less educated. It is worse for older women; those aged eighteen to twenty-nine don't seem to be doing too badly. Women with kids have fared worse than women without kids. The only notable exception to the pattern is black women, who are happier today than they were three decades ago.

There are a number of alternative explanations for these findings. Below is my list, which differs somewhat from the list that Stevenson and Wolfers present:

1. Female happiness was artificially inflated in the 1970s because of the feminist movement and the optimism it engendered. Yes, things have gotten better for women over the last few decades, but

maybe change has happened a lot more slowly than anticipated. Thus, relative to these lofty expectations, things have been a disappointment.

2. Women's lives have become more like men's over the last thirty-five years. Men have historically been less happy than women. So it might not be surprising if the things in the workplace that always made men unhappy are now bedeviling women as well.

3. There was enormous social pressure on women in the old days to pretend they were happy even if they weren't. Now, society allows women to express their feelings openly when they are dissatisfied with life.

4. Related to number three: these self-reported happiness measures are so hopelessly garbled by other factors that they are completely meaningless. The ever-growing army of happiness researchers will go nuts at this suggestion, but there is some pretty good evidence (including a paper by Marianne Bertrand and Sendhil Mullainathan) that declarations of happiness leave a lot to be desired as outcome measures.

Stevenson and Wolfers don't take a stand on what the most likely explanation might be. If I had to wager a guess, I would say numbers three and four are the most plausible.

Meanwhile, I asked a female friend what she thought the answer was, but she was too depressed to respond.

WHAT'S THE BEST ADVICE YOU EVER GOT?
(SJD)

It's that time of year: graduation. Celebrities, dignitaries, and the occasional wild card are ushered forth to send graduates into the future with courage, confidence, conviction (blah blah, blah blah, blah blah) . . .

And then there's a woman we'll call only S., for her mission is a secret one. Her son, N., is about to be graduated from high school, and S. is putting together an "album of advice" for him. She's been writing to all sorts of people (including us) and asking: "What's the best (or worst) advice you've ever been given?" As she writes further: "My mom did this for me when I graduated high school, and I wanted to carry on the tradition for my children. It was the most memorable gift I've ever received."

How could anyone possibly turn down this request? My first inclination was to tell N. that the best advice I could give him was to have a mother who cared enough about her kids to solicit advice from strangers.

Anyway, here's what I sent him. I can't say it's all that interesting, or even such great advice, but this is what came to mind:

Dear N.,

I once received a piece of advice when I was about fourteen that wasn't even meant to be advice, but has stayed with me for my entire life.

I was out fishing on a small lake in a little motorboat with a man named Bernie Duszkiewicz. He was the local barber (well, one of two—but you get the idea: it was a very small town). My father had died when I was ten, and there were a few nice men around town who went out of their way to take me on little adventures. Most of these adventures involved fishing. I didn't really like fishing all that much but I think my mom thought I did, and I was too timid or obedient to ever object.

We were out on the lake, fishing for bass, I suppose, going from one theoretically good spot to the next and catching absolutely nothing. Then it started to rain. Mr. Duszkiewicz drove the boat over toward the shore and anchored us under some low-hanging trees so we wouldn't get drenched. We started casting from there—and lo and behold, I finally caught a fish. It couldn't have been more than six inches long, a sunfish or rock bass, but at least it was a fish. And then I caught another, and another. They were too small to keep but it was fun catching them.

Then the sun came out, and Mr. Duszkiewicz pulled up the anchor. I was a very shy kid and it took everything I had to speak up: "Where are we going? This is a great spot!"

"Ah, we don't want to keep catching these little ones," he said. "They're not worth the time. Let's go catch a real fish."

To be honest, my feelings were a little bit hurt—the fish I was catching *were* real fish, and a lot more fun than catching nothing at all. And we had the same bad luck when we got back out to the deeper spots in the lake: no fish at all.

But the lesson stuck with me. Even though we returned home empty-handed, we went for the big fish. In the short run, this kind of thinking might not be as much fun. But it's the long run you should be thinking about—the big goals, the ones that require a lot of failure along the way. They might be worth it (of course, they might not be, too). It's a lesson in opportunity cost: if you spend all your time catching the little fish, you won't have time—or develop the technique, or the patience—to ever catch the big ones.

Wishing you the very best,

SJD

Well, that's my fish story. The funny thing is that, as memorable as that advice was, I constantly fail to follow it today.

But just think how much worse off I'd be if it weren't at least haunting me, like a second conscience.

THE HIGHEST PRAISE ANYONE COULD EVER GIVE
(SDL)

I got this e-mail from a fan yesterday:

> I read Freakonomics and was—to say the least—
> floored. You are a brilliant thinker and honestly,
> you remind me of me.

Acknowledgments

S uzanne Gluck is our patron saint. Suzanne, we are so grateful for your support and especially your friendship. There are many others at WME to thank as well, including Tracy Fisher, Cathryn Summerhayes, Henry Reisch, Ben Davis, Lori Odierno, Eric Zohn, Dave Wirtschafter, Bradley Singer, and the folks who have over the years propped up everything: Eve Attermann, Erin Malone, Judith Berger, Sarah Ceglarski, Georgia Cool, Caroline Donofrio, Kitty Dulin, Samantha Frank, Evan Goldfried, Mac Hawkins, Christine Price, Clio Seraphim, Mina Shaghaghi, and Liz Tingue.

Huge thanks, as always, to the great crew at William Morrow/HarperCollins, who work so hard on our behalf and on behalf of many other lucky authors. These four Freakonomics books have been a long and wonderful ride with all of you! Special thanks to Henry Ferris, Claire Wachtel, Liate Stehlik, Lisa Gallagher, Michael Morrison, Brian Murray, Jane Friedman, Lynn Grady, Tavia Kowalchuk, Andy Dodds, Dee Dee DeBartlo, Trina Hunn, and the many other talented folks who've contributed so much to this endeavor.

At Penguin UK, we are extremely fortunate to have been edited by a pair of devoted thinkers and good friends, Alexis Kirschbaum and Will Goodlad. Thanks also to Stefan McGrath for continued support.

Thanks also to the wonderful people at the Harry Walker Agency, who routinely send us on great expeditions. And to the Freakonomics Radio crew at WNYC, who do such a great job of turning our rambling into something that approaches coherence.

And then there is the cast of dozens, at least, who have worked so hard over the years on the blog. Truly, it's been a blast.

Thanks to Mary K. Elkins, Lorissa Shepstone and Gordon Clemmons of Being Wicked, and Chad Troutwine and his team for building and constantly rebuilding an online sandbox for us to play in.

At *The New York Times*, thanks especially to Gerry Marzorati, David Shipley, Sasha Koren, Jeremy Zilar, Jason Kleinman, and Brian Ernst.

The blog has had a succession of editors over the years who not only contributed a ton of great writing but also kept the two of us from falling off the wire. Thanks to Rachel Fershleiser, Nicole Tourtelot, Melissa Lafsky, Annika Mengisen, Ryan Hagen, Dwyer Gunn, Mathew Philips, Azure Gilman, Bourree Lam, and Caroline English, with special thanks to Bourree and Dwyer for doing early triage on eight-thousand-plus posts, and to Ryan for, among many other contributions, his piratical Q&A on page 350.

Thanks also to the many guest contributors on the Freakonomics blog over the years, from Q&As to Quorums to occasional essays. We are especially indebted to the awesome cadre of regular contributors, including: Ian Ayres, Captain Steve, Dan Hamermesh, Dean Karlan, Andrew Lo, Sanjoy Mahajan, James McWilliams, Eric Morris, Nathan Myhrvold, Jessica Nagy, Kal Raustiala, Seth Roberts, Steve Sexton, Fred Shapiro, Chris Sprigman, Sudhir Venkatesh, and Justin Wolfers. Special thanks to Captain Steve, James, and Sudhir for letting us put some of their posts in this book.

One component of the blog—one of the best components, to be sure—cannot be conveyed in this book: reader feedback. We've delighted in your clever or insightful or irate comments; your questions and suggestions; your massive curiosity and kindness. Thanks to every single reader: you are what kept us going for ten years.

Notes

CHAPTER 1: WE WERE ONLY TRYING TO HELP

9 **"TERRORISM, PART II"**: "The most hate mail . . . since the abortion-crime story first broke": See *Freakonomics* and John J. Donohue III and Steven D. Levitt, "The Impact of Legalized Abortion on Crime," *The Quarterly Journal of Economics* 116, no. 2 (May 2001). / 11 **"as Gary Becker and Yona Rubinstein have shown . . .":** See Becker and Rubinstein, "Fear and the Response to Terrorism: An Economic Analysis," Centre for Economic Performance Discussion Paper 1079 (Sept. 2011) / 12 **"The work of my University of Chicago colleague Robert Pape suggests . . .":** See, e.g., Pape, *Dying to Win* (Random House, 2005).

13 **"HOW ABOUT A 'WAR ON TAXES?'"**: "David Cay Johnston . . . reports": See Johnston, "I.R.S. Enlists

Help in Collecting Delinquent Taxes," *The New York Times*, August 20, 2006. / 14 **"We touched on this subject in a *Times* column"**: See Dubner and Levitt, "Filling in the Tax Gap," *The New York Times Magazine*, April 2, 2006.

16 **"IF PUBLIC LIBRARIES DIDN'T EXIST . . .":** For further thoughts on this topic, see Dubner, "What I Told the American Library Association," Freakonomics. com, May 5, 2014.

19 **"LET'S JUST GET RID OF TENURE . . .":** See also "The Freakonomics of Tenure," *The Chronicle of Higher Education*, March 23, 2007.

22 **"WHY RESTORING THE MILITARY DRAFT . . .":** "A long report in *Time*": See Mark Thompson, "Restoring the Draft: No Panacea," *Time*, July 21, 2007.

29 **"A FREAKONOMICS PROPOSAL TO HELP . . .":** A "blogger named Noah Smith, who rails on us": See Smith, "Market Priesthood," Noahpinion.com, May 15, 2014.

33 **"AN ALTERNATIVE TO DEMOCRACY?":** "Economists tend to have an indifference towards voting": See Dubner and Levitt, "Why Vote?," *The New York Times Magazine*, November 6, 2005; and Dubner, "We the Sheeple," Freakonomics Radio, October 25, 2012. / 34 **Glen Weyl's voting mechanism:** See Steven P. Lalley and E. Glen Weyl, "Quadratic Voting," SSRN working paper, February 2015. / 35 **"Two other economists . . . have been exploring**

a similar idea": See Jacob K. Goeree and Jingjing Zhang, "Electoral Engineering: One Man, One Vote Bid," working paper, August 27, 2012.

36 **"WOULD PAYING POLITICIANS MORE . . .":** "A research paper by Claudio Ferraz and Frederico Finan": See Ferraz and Finan, "Motivating Politicians: The Impacts of Monetary Incentives on Quality and Performance," NBER working paper, April 2009. / 38 **"Another, more recent paper":** See Finan, Ernest Dal Bó, and Martin Rossi, "Strengthening State Capabilities: The Role of Financial Incentives in the Call to Public Service," *The Quarterly Journal of Economics* 18, no. 3 (April 2013).

CHAPTER 2: LIMBERHAND THE MASTURBATOR AND THE PERILS OF WAYNE

45 **"YOURHIGHNESS MORGAN":** "He sent an *Orlando Sentinel* article": See Joe Williams, "What's in a Name? A Royal Heritage," *Orlando Sentinel*, August 18, 2006. / 46 **"A sad *San Diego Tribune* article":** See "Ex-Navy Marksman Gets 84-to-Life in Gang Shooting," *U-T San Diego*, May 25, 2006.

46 **"WHAT A HEAVENLY NAME":** "Jennifer 8. Lee . . . has the story": See Lee, "And if It's a Boy, Will It Be Lleh?," *The New York Times*, May 18, 2006. / 47 **"The seventieth ranked name":** a great resource for baby-naming trends can be found on the Social

Security Administration's website: http://www.ssa.
gov/oact/babynames/.

47 "THE UNPREDICTABILITY OF BABY NAMES": See
"Hurricane Dealt Blow to Popularity of Katrina
as Baby Name," *The New York Times* (Associated
Press article), May 13, 2007; and, again, a good
resource for baby-name trends in general is at http://
www.ssa.gov/oact/babynames/.

49 "BEAT THIS APTONYM": "LIMBERHAND THE MASTUR-
BATOR": See *State of Idaho v. Dale D. Limberhand*,
No. 17656, Court of Appeals of Idaho, March 14,
1990.

CHAPTER 3: HURRAY FOR HIGH GAS PRICES!

57 "IF CRACK DEALERS TOOK LESSONS . . .": "a TV
news report . . .": See Eileen Faxas, "Up Close:
Cost of Generic Drugs Varies Widely," KHOU-TV.
com, December 13, 2003. / 59 **An extensive price
comparison**": See "Generic Prescription Drug
Price Comparison Chart," WXYZ-TV.com. / 59 **"A
Consumer Reports survey"**: See "Generic Drugs:
Shop Around for the Best Deals," ConsumerReports.
org. / 59 **"A research report . . . Dianne Feinstein"**:
See "Senator Feinstein Urges Californians to Be
Aware That Generic Drug Prices Vary Greatly
From Pharmacy to Pharmacy," May 8, 2006. / 59
"A comprehensive Wall Street Journal article":
Sarah Rubenstein, "Why Generic Doesn't Always

Mean Cheap," *The Wall Street Journal,* March 13, 2007.

63 "FOR $25 MILLION, NO WAY . . .": "The virtues of offering big prizes to encourage . . . curing disease": See Levitt, "Fight Global Pandemics (or at Least Find a Good Excuse When You're Playing Hooky)," Freakonomics.com, May 18, 2007; "or improving Netflix's algorithms": See Levitt, "Netflix $ Million Prize," Freakonomics.com, October 6, 2006. / 65 **"As reported by ABC News":** See Matthew Cole, "U.S. Will Not Pay $25 Million Osama Bin Laden Reward, Officials Say," ABCNews.com, May 19, 2011.

67 "CAN WE PLEASE GET RID OF THE PENNY ALREADY?": A "*60 Minutes* segment called 'Making Cents'": See Morley Safer, "Should We Make Cents?," *60 Minutes,* February 10, 2008.

78 "JANE SIBERRY SNAPS": "Anybody remember when Levitt announced . . .": See Levitt, "The Two Smartest Musicians I Ever Met," Freakonomics.com, April 5, 2006; and Levitt, "From Now on I Will Leave the Reporting to Dubner," Freakonomics. com, April 9, 2006.

79 "HOW MUCH TAX ARE ATHLETES . . .": "Manny Pacquiao will probably never fight in New York": See "Manny Pacquiao Won't Ever Fight in New York Due to State Tax Rates," *The Wall Street Journal,* August 7, 2013. / 80 **"Pacquiao may never fight**

anywhere in the U.S. again": See Lance Pugmire, "Promoter: Manny Pacquiao May Never Again Fight in the U.S.," *The Los Angeles Times*, May 31, 2013. / 81 **"Phil Mickelson . . . 'going to have to make some drastic changes'"**: See "Golfer Phil Mickelson Plans 'Drastic Changes' Over Taxes," CBSNews.com, January 21, 2013. / 82 **"In *Forbes*, Kurt Badenhausen wrote . . . about Mickelson's British tax tab"**: See Badenhausen, "Phil Mickelson Wins Historic British Open and Incurs 61% Tax Rate," Forbes.com, July 22, 2013. / 82 **"Mick Jagger fled the U.K."**: See Larry King interview with Jagger on *Larry King Live*, CNN, May 18, 2010.

95 **"HURRAY FOR HIGH GAS PRICES!"**: historic gas prices are drawn from the U.S. Energy Information Administration; see also AAA's fuelgaugereport.com. / 97 **"In a paper I was proud to publish"**: See Aaron S. Edlin and Pinar Karaca Mandic, "The Accident Externality From Driving," *The Journal of Political Economy* 114.5 (2006). / 98 **"According to a National Academy of Sciences report"**: See *Tires and Passenger Vehicle Fuel Economy: Informing Consumers, Improving Performance*, The National Academies Press, Special Report 286 (2006). / 99 **Expensive gas leads to more motorcycle fatalities**: See He Zhu, Fernando A. Wilson, and Jim P. Stimpson, "The Relationship Between

Gasoline Price and Patterns of Motorcycle Fatalities and Injuries," *Injury Prevention* (2014).

CHAPTER 4: CONTESTED

106 "CONTEST: a six-word motto . . .": "England's reluctant search for a national motto": See Sarah Lyall, "Britain Seeks Its Essence, and Finds Punch Lines," *The New York Times*, January 26, 2008. / 106 **"A new book on six-word memoirs"**: See Rachel Fershleiser and Larry Smith (eds.), *Not Quite What I Was Planning: Six-Word Memoirs by Writers Famous and Obscure* (HarperCollins, 2008). Side note: Rachel Fershleiser was the first editor of the Freakonomics blog.

CHAPTER 5: HOW TO BE SCARED OF THE WRONG THING

110 "WHOA NELLIE": "A 1990 CDC report": See "Current Trends Injuries Associated with Horseback Riding—United States, 1987 and 1988," Centers for Disease Control. / 111 **Horseback riders "are often under the influence of alcohol"**: See "Alcohol Use and Horseback-Riding-Associated Fatalities—North Carolina, 1979–1989," Centers for Disease Control.

112 "WHAT THE SECRETARY OF TRANSPORTATION . . .": "On his official government blog": See Ray LaHood, "Current Data Makes It Clear: Child Safety Seas and Booster Seats Save Lives, Prevent Injury," Fast Lane

(U.S. Dept. of Transportation blog), October 22, 2009. / 112 **"My research on child safety seats"**: See Levitt and Dubner, *SuperFreakonomics* (William Morrow, 2009); and Dubner and Levitt, "The Seat-Belt Solution," *The New York Times Magazine*, July 10, 2005. / 113 **"When I first told him about my work on teacher cheating"**: See Levitt and Dubner, *Freakonomics* (William Morrow, 2005).

119 **" 'PEAK OIL' "**: "A recent . . . cover story": See Peter Maass, "The Breaking Point," *The New York Times Magazine*, August 21, 2005.

124 **"BETTING ON PEAK OIL"**: "John Tierney wrote a great . . . column": See Tierney, "The $10,000 Question," *The New York Times*, August 23, 2005. / 127 **"Sadly, Matthew Simmons died"**: See Tierney, "Economic Optimism? Yes, I'll Take That Bet," *The New York Times*, December 27, 2010.

127 **"DOES OBESITY KILL?"**: "An interesting paper": See Shin-Yi Chou, Michael Grossman, and Henry Saffer, "An Economic Analysis of Adult Obesity: Results from the Behavioral Risk Factor Surveillance System," NBER working paper No. 9247, October 2002. / 128 **"a paper calling into doubt"**: See Jonathan Gruber and Michael Frakes, "Does Falling Smoking Lead to Rising Obesity?," NBER working paper No. 11483, July 2005. / 129 **"The panic over obesity may be as big a problem . . ."**: See J. Eric Oliver, *Fat Politics: The Real Story Behind*

America's Obesity Epidemic (Oxford University Press, 2006). / 130 **"The tour company had been using the old standard . . .":** Al Baker and Matthew L. Wald, "Weight Rules for Passengers Called Obsolete in Capsizing," *The New York Times*, July 1, 2006.

131 **"DANIEL KAHNEMAN ANSWERS YOUR QUESTIONS":** See Kahneman, *Thinking, Fast and Slow* (Farrar, Straus and Giroux, 2011).

140 **"FOUR REASONS WHY THE U.S. CRACKDOWN . . .":** "The U.S. government recently shut down . . .": See Matt Richtel, "U.S. Cracks Down on Online Gambling," *The New York Times*, April 15, 2011. / 143 **"I recently co-wrote a paper":** Levitt and Thomas J. Miles, "The Role of Skill Versus Luck in Poker," NBER working paper 17023, May 2011.

144 **"THE COST OF FEARING STRANGERS":** "an AirTran spokesman told *The Washington Post*": Amy Gardner, "9 Muslim Passengers Removed From Jet," *The Washington Post*, January 2, 2009. / 147 **"How about child abduction? . . . a 2007 *Slate* article explains":** Christopher Beam, "800,000 Missing Kids? Really?," *Slate.com*, January 17, 2007.

CHAPTER 6: IF YOU'RE NOT CHEATING, YOU'RE NOT TRYING

148 **"CHEATING TO BE HOT":** "like the office workers who put money . . .": See Levitt and Dubner,

Freakonomics (William Morrow, 2005). / 149 **"Farhad Manjoo's article . . . about a contest"**: Manjoo, "How Bots Rigged D.C.'s 'Hot' Reporter Contest," Salon.com, August 22, 2007. / 150 **"We have been accused of stuffing a ballot box"**: See Melissa Lafsky, "*Freakonomics* v. *Lolita*: Can You Tell the Difference?," Freakonomics.com, June 18, 2007.

150 **"WHY DO YOU LIE?"**: "A new paper by César Martinelli and Susan W. Parker": See Martinelli and Parker, "Deception and Misreporting in a Social Program," Centro de Investigacion Economica discussion paper 06-02, June 2006. / 153 **"An article about the lack of hand hygiene in hospitals"**: See Dubner and Levitt, "Selling Soap," *The New York Times Magazine*, September 24, 2006. / 154 **"the topics that online daters are most likely to lie about" and the "risky business of election polling"**: See Levitt and Dubner, *Freakonomics* (William Morrow, 2005).

154 **"HOW TO CHEAT THE MUMBAI TRAIN SYSTEM"**: "A blogger named Ganesh Kulkarni": See Kulkarni, "What a Business Model!," ganeshayan.blogspot.com, March 21, 2007.

167 **"HOW WE WOULD FIGHT STEROIDS . . ."**: "Zelinsky . . . has proposed": See Aaron Zelinsky, "Put More Muscle in Baseball Drug Tests," *The Hartford Courant*, December 18, 2007.

169 "HOW NOT TO CHEAT": "within a few days, they were discovered": See adanthar, "Beat: Absolute is *actually* rigged (serious) (read me)," September 15, 2007, twoplustwo.com.

171 "THE ABSOLUTE POKER CHEATING SCANDAL . . .": "*The Washington Post* has followed up": See Gilbert M. Gaul, "Cheating Scandals Raise New Questions About Honesty, Security of Internet Gambling," *The Washington Post*, November 30, 2008. / 174 **"Update"**: See Gaul, "Timeline: Catching the Cheaters," *The Washington Post*.

174 "TAX CHEATS OR TAX IDIOTS?": "We once wrote a column about tax cheating": Dubner and Levitt, "Filling in the Tax Gap," *The New York Times Magazine*, April 2, 2006. / 176 **"The Simple Return"**: See Austan Goolsbee, "The Simple Return: Reducing America's Tax Burden Through Return-Free Filing," The Hamilton Project discussion paper 2006-04, July 2006.

176 "HAVE D.C.'S 'BEST SCHOOLS' BEEN CHEATING?": "A *USA Today* investigation": See Jack Gillum and Marisol Bello, "When Standardized Test Scores Soared in D.C., Were the Gains Real?," *USA Today*, March 30, 2011. / 177 **"Kaya Henderson did request a review"**: See Gillum, Bello, and Scott Elliott, "D.C. to Dig Deeper on Test Score Irregularities," *USA Today*, March 30, 2011. / 177 **"When Brian Jacob and I investigated teacher**

cheating": See Levitt and Dubner, *Freakonomics* (William Morrow, 2005); and Brian A. Jacob and Levitt, "Rotten Apples: An Investigation of the Prevalence and Predictors of Teacher Cheating," *The Quarterly Journal of Economics* (August 2003).

CHAPTER 7: BUT IS IT GOOD FOR THE PLANET?

182 **"IS THE ENDANGERED SPECIES ACT . . .":** "He's got a new working paper": See John A. List, Michael Margolis, and Daniel E. Osgood, "Is the Endangered Species Act Endangering Species?," NBER working paper 12777, December 2006. / 183 **"Sam Peltzman's observation that only 39 of the 1,3000 species list have ever been removed":** See Sam Peltzman, "Regulation and the Natural Progress of Opulence," American Enterprise Institute monograph, May 23, 2005.

184 **"BE GREEN: Drive":** "via John Tierney's blog": John Tierney, "How Virtuous Is Ed Begley Jr.?," *The New York Times* (TierneyLab), February 25, 2008. / 184 **"Goodall is no right-wing nut":** See Chris Goodall, *How to Live a Low-Carbon Life* (Earthscan, 2007).

185 **"DO WE REALLY NEED A FEW BILLION LOCAVORES?":** "As we have written before": See Dubner and Levitt, "Laid-Back Labor," *The New York Times Magazine*, May 6, 2007. / 189 **"consider the 'food**

miles' argument and a recent article": See
Christopher L. Weber and H. Scott Matthews,
"Food-Miles and the Relative Climate Impacts of
Food Choices in the United States," *Environmental
Science & Technology* 42, no. 10 (April 2008).

190 "GOING GREEN TO INCREASE PROFITS": "As Mary
MacPherson Lane writes in an A.P. article": See
Mary MacPherson Lane, "Brothel Cuts Rates for
'Green' Customers," Associated Press, October 17,
2009.

194 "HOW ABOUT THEM (WRAPPED) APPLES": "Similar
numbers have been found for potatoes and grapes":
See "Food Packaging and Climate Change," car-
boncommentary.com, October 29, 2007. / 196 **"One
study estimates that U.S. consumers throw out
about half the food they buy"**: See J. Lundqvist, C.
de Fraiture, and D. Molden, "Saving Water: From
Field to Fork—Curbing Losses and Wastage in the
Food Chain," SIWI policy brief (2008).

198 "AGNOSTIC CARNIVORES AND GLOBAL WARMING . . .":
"Neither he nor 350.org will actively promote a
vegan diet": when asked in February, 2015, if this
were still the case, a 350.org spokesperson replied:
"No, we still don't have an active campaign pushing
for people to go vegan. Then again, we don't have
active campaigns for people to drive less, recycle,
use less paper, or any other long list of worthy ways
to combat climate change. 350.org doesn't work on

individual lifestyle change—there are lots of good groups that do that—but instead works at building more of a social movement to fight the problem." / 198 **"A recent report from the World Preservation Foundation confirms"**: See "Reducing Shorter-Lived Climate Forcers Through Dietary Change," World Preservation Foundation. / 200 **"A recent article McKibben wrote for *Orion*"**: See Bill McKibben, "The Only Way to Have a Cow," *Orion*, April 2010.

204 **"HEY BABY, IS THAT A PRIUS YOU'RE DRIVING?"**: "A research paper written by Alison and Steve Sexton": See "Conspicuous Conservation: The Prius Effect and Willingness to Pay for Environmental Bona Fides," working paper, June 30, 2011. / 205 **"Here's how Steve Sexton explains it"**: See Stephen J. Dubner, "Hey Baby, Is That a Prius You're Driving?," Freakonomics Radio, July 7, 2011.

CHAPTER 8: HIT ON 21

208 **"VEGAS RULES"**: "for a *Times* column on Super Bowl gambling": See Dubner and Levitt, "Dissecting the Line," *The New York Times Magazine*, February 5, 2006.

212 **"ONE CARD AWAY . . ."**: "Brandon Adams . . . a great writer": See Adams, *Broke: A Poker Novel* (iUniverse, 2006).

219 "WHAT ARE MY CHANCES OF MAKING THE CHAM-
PIONS TOUR...": "My friend Anders Ericsson
popularized the magic number of 10,000 hours of
practice": See Dubner and Levitt, "A Star is Made,"
The New York Times Magazine, May 7, 2006; K.
Anders Ericsson, Neil Charness, Paul J. Feltovich,
and Robert R. Hoffman, *The Cambridge Handbook
of Expertise and Expert Performance*, Cambridge
University Press, 2006.

228 "LOSS AVERSION IN THE NFL": "Just about every-
one... capuchin monkeys": See Dubner and
Levitt, "Monkey Business," *The New York Times
Magazine*, June 5, 2005.

230 "BILL BELICHICK IS GREAT": "Teams seem to punt
way too much": See David Romer, "Do Firms
Maximize? Evidence from Professional Football,"
Journal of Political Economy 118, no. 2 (2006). / 231
**"I've seen the same thing in my research on pen-
alty kicks in soccer":** Pierre-André Chiappori,
Steven D. Levitt, and Timothy Groseclose, "Test-
ing Mixed-Strategy Equilibria When Players Are
Heterogeneous: The Case of Penalty Kicks in
Soccer," *The American Economic Review* 92, no. 4
(September 2002).

231 "HOW ADVANTAGEOUS IS HOME-FIELD ADVAN-
TAGE...": See Tobias Moskowitz and L. Jon
Wertheim, *Scorecasting: The Hidden Influences
Behind How Sports Are Played and Games Are Won*

(Crown Archetype, 2011). / 232 **"Levitt once wrote an academic paper about ... home underdogs"**: See Levitt, "Why Are Gambling Markets Organised So Differently From Financial Markets?," *The Economic Journal* 114 (April, 2004). / 232 **"which we wrote about further in the *Times*"**: See Dubner and Levitt, "Dissecting the Line," *The New York Times Magazine*, February 5, 2006. / 234 **"a research paper ... about home-field advantage in the Bundesliga"**: See Thomas J. Dohmen, "In Support of the Supporters? Do Social Forces Shape Decisions of the Impartial?," IZA discussion paper No. 755, April 2003.

235 **"TEN REASONS TO LIKE THE PITTSBURGH STEELERS"**: "A voice that sounded like gravel and Yiddish tossed in a blender": See Myron Cope, *Double Yoi!* (Sports Publishing, 2002). / 239 **"Franco Harris ... yours truly even wrote a book about his strange appeal"**: See Stephen J. Dubner, *Confessions of a Hero-Worshiper* (William Morrow, 2003). / 241 **"Paucity of great books about football"**: See Roy Blount Jr., *About Three Bricks Shy of a Load* (Little, Brown and Company, 1974).

CHAPTER 9: WHEN TO ROB A BANK

248 **"WHEN TO ROB A BANK"**: "a story I was told while visiting Iowa": See "Burnice Comes Home," *Time*, July 8, 1966. / 250 **"According to the FBI"**: see the

FBI's National Incident-Based Reporting System (NIBRS) series. / 251 **"a trove of robbery data from the British Bankers' Association":** see Barry Reilly, Neil Rickman, and Robert Witt, "Robbing Banks: Crime Does Pay—But Not Very Much," *Significance* (The Royal Statistical Society, June 2012).

253 "DON'T REMIND CRIMINALS THEY ARE CRIMINALS": "failed to replicate them in one study I did": See Roland G. Fryer, Steven D. Levitt, and John A. List, "Exploring the Impact of Financial Incentives on Stereotype Threat: Evidence From a Pilot Study," *American Economic Review: Papers & Proceedings* 98, no. 2 (2008). / 254 **"In an interesting new study:** see Alain Cohn, Michel André Maréchal, and Thomas Noll "Bad Boys: The Effect of Criminal Identity on Dishonesty," University of Zurich working paper No. 132 (October 2013).

264 "DON'T BURN THE FOOD": "In a sample of 13 African countries between 1999 and 2004": The Demographic and Health Surveys Program, U.S. Agency for International Development.

266 "IS PLAXICO BURRESS AN ANOMALY?": "A few years back, I wrote an article": See Dubner, "Life Is a Contact Sport," *The New York Times Magazine*, August 18, 2002. / 268 **"According to an ESPN report":** See Arty Berko, Steve Delsohn, and Lindsay Rovegno, "Athletes and Guns," *Outside the Lines* and ESPN.com, December 15, 2006.

273 "**WHAT'S THE BEST WAY TO CUT GUN DEATHS?**": "For a project . . . I conducted": See Philip J. Cook, Jens Ludwig, Sudhir Venkatesh, and Anthony A. Braga, "Underground Gun Markets," *The Economic Journal* 117, no. 524 (November 2007).

282 "**WEIRD BUT TRUE . . .**": "Here's the news": See Nellie Andreeva, "NBC Buys 'Freakonomics'-Inspired Drama Procedural Produced by Kelsey Grammer, *Deadline.com*, August 7, 2012.

CHAPTER 10: MORE SEX PLEASE, WE'RE ECONOMISTS

285 "**BREAKING NEWS . . .**": World Cup brothel boom "hasn't happened at all": See, e.g., Mark Landler, "World Cup Brings Little Pleasure to German Brothels," *The New York Times*, July 3, 2006.

286 "**AN IMMODEST PROPOSAL: TIME FOR A SEX TAX?**": "Bernard Gladstone proposed such a measure in his state": See "The Nation: Sex Tax," Time, January 25, 1971. / 289 **"The one tax that would probably be overpaid"**: See "Sex Tax: 'Broad-Based,'" *The Tech* (MIT newspaper), January 13, 1971.

290 "**MORE SEX PLEASE, WE'RE ECONOMISTS**": "Women choke under pressure": See Steven E. Landsburg, "Women Are Chokers," Slate.com, February 9, 2007. / 290 **"Miserliness is a form of generosity"**: See Landsburg, "What I Like About Scrooge," Slate.com, December 20, 2006. / 290 **"He is the author of the books"**: See e.g., Landsburg, *The Armchair*

Economist (Free Press, 1993); Landsburg, *Fair Play* (Free Press, 1997); Landsburg, *More Sex Is Safer Sex* (Free Press, 2007).

292 "I'M A HIGH-END CALL GIRL; ASK ME ANYTHING": for a fuller treatment of Allie's business, see Levitt and Dubner, *SuperFreakonomics* (William Morrow, 2009). Allie was also featured in Dubner, "The Upside of Quitting," Freakonomics Radio, September 30, 2011.

300 "FREAKONOMICS RADIO GETS RESULTS": "How drivers are legally allowed to fatally run over pedestrians": See Dubner, "The Most Dangerous Machine," Freakonomics Radio, May 1, 2014. / 300 **"Fighting Poverty With Actual Evidence":** See Dubner, "Fighting Poverty With Actual Evidence," Freakonomics Radio, November 27, 2013. / 300 **"How avocadoes help fund Mexican crime cartels":** See Dubner, "What Came First, the Chicken or the Avocado?," Freakonomics Radio, April 24, 2014. / 300 **"What You Don't Know About Online Dating":** See Dubner, "What You Don't Know About Online Dating," Freakonomics Radio, February 6, 2014; this episode featured the research of Stanford economist Paul Oyer, author of *Everything I Ever Needed to Know About Economics I Learned from Online Dating* (Harvard Business Review Press, 2014).

CHAPTER 11: KALEIDOSCOPIA

315 **"IF YOU LIKE HOAXES"**: "You have to admit that this is a pretty good one": See Sarah Lyall, "In Literary London, the Strange Case of the Steamy Letter," *The New York Times*, August 31, 2006.

316 **"FROM GOOD TO GREAT . . . TO BELOW AVERAGE"**: See Jim Collins, *Good to Great: Why Some Companies Make the Leap . . . and Others Don't* (HarperCollins, 2001). / 317 **"The classic Peters and Waterman"**: See Thomas J. Peters and Robert H. Waterman, Jr., *In Search of Excellence: Lessons from America's Best-Run Companies* (Harper & Row, 1982; HarperBusiness Essentials, 2004).

320 **"WHY I LIKE WRITING ABOUT ECONOMISTS"**: "My mother had an extraordinary (and long-buried) story to tell": See Dubner, *Turbulent Souls: A Catholic Son's Return to His Jewish Family* (William Morrow, 1998); republished as *Choosing My Religion: A Memoir of a Family Beyond Belief* (HarperPerennial, 2006.) / 320 **"I've interviewed Ted Kaczynski, the Unabomber"**: See Dubner, "I Don't Want to Live Long. I Would Rather Get the Death Penalty Than Spend the Rest of My Life in Prison," *Time*, October 18, 1999. / 320 **"The rookie class of the N.F.L."**: See Dubner, "Life Is a Contact Sport," *The New York Times Magazine*, August 18, 2002. / 320 **"A remarkable cat burglar

who stole only sterling silver": See Dubner, "The Silver Thief," *The New Yorker*, May 17, 2004. / 321 **"After I wrote about the economist Roland Fryer"**: See Dubner, "Toward a Unified Theory of Black America," *The New York Times Magazine*, March 20, 2005.

CHAPTER 12: WHEN YOU'RE A JET . . .

350 **"PIRATE ECONOMICS 101"**: Ryan Hagen was at the time of this Q&A a Freakonomics research assistant who contributed mightily to the blog, the books, and elsewhere; he is currently completing his Ph.D. in sociology at Columbia. / 350 **"The *Maersk* crew has returned home"**: See Matt Zapotosky, "Amid Breakfast of Champions, Pirated Ship's Crew Shares a Story of Turnabout," *The Washington Post*, April 17, 2009. / 350 **"But with tensions growing"**: See Reuters, "Pirates Attack U.S. Ship Off Somalia," *The New York Times*, April 14, 2009. / 350 **"*The Invisible Hook*"**: See Peter T. Leeson, *The Invisible Hook: The Hidden Economics of Pirates* (Princeton, 2009).

355 **"THE VISIBLE HAND"**: "Some new research": See Jennifer L. Doleac and Luke C.D. Stein, "The Visible Hand: Race and Online Market Outcomes," SSRN working paper, May 1, 2010.

361 **"HOW PURE IS YOUR ALTRUISM"**: "Consider a recent paper": See Philip H. Brown and Jessica H. Minty,

"Media Coverage and Charitable Giving After the 2004 Tsunami," William Davidson Institute working paper No. 855, December 2006. / 365 **"Considering how unevenly disaster aid is often distributed"**: See, e.g., "Tsunami Aid 'Went to the Richest," BBC.com, June 25, 2005. / 365 **"The single best thing . . . is to be an attractive blond woman"**: See Craig E. Landry, Andreas Lange, John A. List, Michael K. Price, and Nicholas G. Rupp, "Toward an Understanding of the Economics of Charity: Evidence from a Field Experiment," *Quarterly Journal of Economics* 121, no. 2 (May 2006).

376 **"BRIBING KIDS TO TRY HARDER ON TESTS"**: "We recently wrote up the results": See Steven D. Levitt, John A. List, Susanne Neckermann, and Sally Sadoff, "The Impact of Short-Term Incentives on Student Performance," University of Chicago working paper, September 2011.

380 **"SHRIMPONOMICS"**: "According to one academic article, the real price of shrimp fell": See U. Rashid Sumaila, A. Dale Marsden, Reg Watson, and Daniel Pauly, "A Global Ex-Vessel Fish Price Database: Construction and Applications," *Journal of Bioeconomics* 9, no. 1 (April, 2007). / 383 **"An article in *Slate* argues . . ."**: Brendan Koerner, "The Shrimp Factor," Slate.com, January 13, 2006.

384 "WHY ARE WOMEN SO UNHAPPY?": "The first was in the form of a *Times* op-ed": See Betsey Stevenson and Justin Wolfers, "Divorced From Reality," *The New York Times*, September 29, 2007. / 384 **"Stevenson and Wolfers released a new study":** See Stevenson and Wolfers, "The Paradox of Declining Female Happiness," IZA discussion paper No. 42347 (2009). / 386 **"There is some pretty good evidence . . . that declarations of happiness leave a lot to be desired":** See Marianne Bertrand and Sendhil Mullainathan, "Do People Mean What They Say? Implications for Subjective Survey Data," MIT Economics working paper No. 01-04 (January 2001).

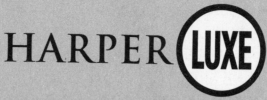

HARPER LUXE

THE NEW LUXURY IN READING

We hope you enjoyed reading
our new, comfortable print size and found it
an experience you would like to repeat.

Well – you're in luck!

HarperLuxe offers the finest in fiction and
nonfiction books in this same larger print size and
paperback format. Light and easy to read, HarperLuxe
paperbacks are for book lovers who want to see
what they are reading without the strain.

For a full listing of titles and
new releases to come, please visit our website:

www.HarperLuxe.com

SEEING IS BELIEVING!

JUN - 9 2015